DIET AND BREAST CANCER

ADVANCES IN EXPERIMENTAL MEDICINE AND BIOLOGY

Recent Volumes in this Series

Volume 357
LACTOFERRIN: Structure and Function
Edited by T. William Hutchens, Sylvia V. Rumball, and Bo Lönnerdal

Volume 358
ACTIN: Biophysics, Biochemistry, and Cell Biology
Edited by James E. Estes and Paul J. Higgins

Volume 359
TAURINE IN HEALTH AND DISEASE
Edited by Ryan J. Huxtable and Dietrich Michalk

Volume 360
ARTERIAL CHEMORECEPTORS: Cell to System
Edited by Ronan G. O'Regan, Philip Nolan, Daniel S. McQueen, and David J. Paterson

Volume 361
OXYGEN TRANSPORT TO TISSUE XVI
Edited by Michael C. Hogan, Odile Mathieu-Costello, David C. Poole,
and Peter D. Wagner

Volume 362
ASPARTIC PROTEINASES: Structure, Function, Biology, and Biomedical Implications
Edited by Kanji Takahashi

Volume 363
NEUROCHEMISTRY IN CLINICAL APPLICATION
Edited by Lily C. Tang and Steven J. Tang

Volume 364
DIET AND BREAST CANCER
Edited under the auspices of the American Institute for Cancer Research
Scientific Editor: Elizabeth K. Weisburger

Volume 365
MECHANISMS OF LYMPHOCYTE ACTIVATION AND IMMUNE REGULATION V:
Molecular Basis of Signal Transduction
Edited by Sudhir Gupta, William E. Paul, Anthony DeFranco, and Roger Perlmutter

DIET AND BREAST CANCER

Edited under the auspices of the American Institute for Cancer Research

Scientific editor

Elizabeth K. Weisburger

American Institute for Cancer Research
Washington, D.C.

SPRINGER SCIENCE+BUSINESS MEDIA, LLC

Library of Congress Cataloging-in-Publication Data

On file

Proceedings of the American Institute for Cancer Research's Fourth Annual Conference on Diet and Cancer: Diet and Breast Cancer, held September 2–3, 1993, in Washington, D.C.

ISBN 978-1-4613-6068-1   ISBN 978-1-4615-2510-3 (eBook)
DOI 10.1007/978-1-4615-2510-3

Preface

The fourth annual American Institute for Cancer Research (AICR) conference on diet, nutrition and cancer was held at the L'Enfant Plaza Hotel in Washington, D.C., September 2-3, 1993. In keeping with present concerns and in line with current trends, the theme was "Diet and Breast Cancer." This proceedings volume is comprised of chapters from the platform presentations of the two day conference and abstracts from the poster session held at the end of the first day.

Experimentally, there is sufficient evidence to support a relationship between dietary fat and the risk of breast cancer. A meta-analysis was provided by data from 114 experiments with over 10,000 animals, divided into groups fed *ad libitum* on diets with different levels or sources of fat, or different levels of energy restriction. This exercise suggested that linoleic acid was a major determinant of mammary tumor development but that other fatty acids also enhanced mammary tumor development in animals. However, as mentioned by several speakers, results from epidemiological studies often are conflicting, thus leading to confusion among both health professionals and the public. Surveys of specific populations which have migrated from countries with low breast cancer rates to those with higher rates are often some of the most compelling studies with respect to a high fat diet–breast cancer association. Nonetheless, various cohort and prospective studies, some quite large, did not appear to show a relationship between consumption of fat (any type) and breast cancer. Overall, the relative risk was 1; in some cases there even was a negative association. In contrast, for colon cancer, there was a positive association between the level of dietary fat and disease incidence. However, a meta-analysis of 12 case-control studies pointed toward a positive association between fat intake and breast cancer for postmenopausal women. In contrast, the data from the Nurses' Health Study showed no evidence to support a dietary fat–breast cancer risk hypothesis. Again this same cohort demonstrated a positive association between total fat and animal fat and colon cancer.

Against this background of conflicting evidence, there seems to be agreement on one point. For women who already have breast cancer, proper clinical management of their cases should prudently include a diet with fat levels of 25% or lower. Clinical data supporting an association between reduction in dietary fat and better survival for breast cancer patients come from: observations in patients from different areas with differing cancer rates, obesity at time of resection, weight gain after resection and possibly the level of dietary fat at time of resection. Dietary fat may influence breast cancer by leading to changes in circulatory hormone levels, to changes in cell membrane structure and changes in the host immune-modulating cells, in addition to changes in the breast structure itself and changes in blood prostaglandin, cytokines, and altered regulatory gene expression.

A mechanistic basis for the poorer outcome in breast cancer patients who are obese or gain weight after therapy was provided by studies on the diet-sex hormone interaction. Vegetarians had lower plasma estrone and estradiol levels than did women who were omnivores; the estradiol level could be correlated with dietary linoleic acid, a likely constituent of dietary fat. Thus, reducing fat intake from 35% to 21% of the diet in women with cystic breast disease led to appreciable declines in serum estrone, estradiol and total estrogen. Furthermore, besides being low in fat, vegetarian based diets are generally higher in fiber, thus decreasing enterohepatic circulation of estrogens and increasing fecal excretion. Another factor is that fiber contains lignans which may bind weakly to type I estrogen receptors. Controlled metabolic studies in women corroborated that a lower fat, high fiber diet decreased serum concentrations of estrone (as the sulfate), of estradiol and of sex hormone binding globulin. Thus, it was surmised that by changing to a lower fat, high fiber diet, declines in the order of 16-24% in breast cancer incidence could be achieved.

The most convincing evidence on the role of dietary fat in breast cancer is derived from animal studies. For various rodent mammary tumor models using different types of chemical carcinogens, spontaneous tumors, and transplanted tumors have confirmed that high dietary fat has a unique ability to enhance or promote mammary tumor development. Although the effect was independent of increased calorie intake, a few other types of tumors also grew faster in animals on a high-fat diet. Further investigation has implicated unsaturated free fatty acids and polyunsaturated fatty acids as responsible for this effect. Monounsaturated and saturated fats were not active. Fish oils, especially those containing omega-3 polyunsaturated fatty acids reduced tumor incidence after carcinogen treatment and lengthened the latent period. Since serum prolactin, progesterone and estradiol levels were similar in rats on high fat and normal diets, some explanation other than that of hormones alone was indicated. One explanation was that the dietary lipids may modify the physicochemical environment of membranes, of hormone receptors and/or enzymes of the tumor cells. Another was that dietary lipid may alter the pool of precursor molecules available for eicosanoid metabolism, involving both cyclooxygenase and lipoxygenase pathways.

Eicosanoids also modulate immune function, another possible mechanism for the effect of the polyunsaturated fats. A more detailed consideration of the alteration in immune function by dietary fat discussed the metabolic pathway of fatty acids through desaturation and elongation to yield arachidonic, eicosatrienoic and eicosapentaenoic fatty acids. Arachidonic acid can be metabolized through either a cyclooxygenase or lipoxygenase pathway leading to prostaglandins or thromboxane and prostacyclin, which modulate macrophage function and thus the immune system. The ability of macrophages to destroy tumor cells depends on the secretion of various lytic substances by the macrophage; in turn the fatty acids delivered to the macrophage and incorporated into its structure determine its physiological activity. Lipopolysaccharide appears to be a signal to kill tumor cells, after the macrophages had been primed by interferon gamma. However, several stimulating and suppressive agents may function in a network. Macrophages are also a principal source of tumor necrosis factor alpha (TNFα) or cachectin. Lipopolysaccharide also induces protein phosphorylation by protein kinase C. Dietary fat may be associated with an alteration in gene regulation, specifically an antiproliferative gene.

The special role of linoleic acid in breast cancer, both as an inhibitory factor and an enhancing factor, was discussed by several groups of authors. Growth of human breast cancer cells in culture was stimulated by linoleic acid; the results were carried over to a nude mouse metastatic model (growth in mammary fat pads). Higher levels of dietary linoleic acid stimulated incidence and metastatic burden but fish oil (omega-3) decreased these effects. On the other hand, conjugated linoleic acid (CLA), a mixture of positional and geometric isomers of linoleic acid had a suppressing effect on skin carcinogenesis induced by DMBA and promoted by TPA (12-O-tetradecanoylphorbol-13-acetate). Similarly, CLA reduced forestomach tumors induced by benzo[a]pyrene in mice, mammary tumors in female rats given DMBA, but not intestinal tumors induced in Fischer rats by azoxymethane. Reinvestigation showed dietary administration of CLA (at a level of 0.1%) for 5 weeks had an anticarcinogenic effect in either the rat-DMBA or rat-NMU model. Mechanistic studies showed CLA inhibited ornithine decarboxylase, a marker for tumor promotion. However, further research on the mechanism of action is needed.

Evidence for the chemopreventive action of vitamin A and synthetic retinoids on both the DMBA and NMU rat mammary tumor model was reviewed. A combination of retinoid, especially N-(4-hydroxyphenyl) retinamide (4-HPR) and other agents such as tamoxifen or dehydroepiandrosterone was more effective than either agent alone.

Despite the effectiveness of retinoids in animal experiments, only a few are being evaluated as chemopreventive agents in humans, notably retinoic acid, retinol and beta-carotene. However, it is not clear that dietary supplementation of retinoids in groups with adequate levels of vitamin A will increase serum levels to preventive levels.

Vitamin D was also an important player, with two speakers discussing both experimental and human effects. The hypothesis given was that low levels of vitamin D and dietary calcium, which are biochemically interrelated, increase the promoting effect of high dietary fat on chemically-induced mammary cancer in animals, that current calcium and vitamin D dietary intake in the U.S. is below RDA (recommended dietary allowance) and that reduction of breast cancer risk might be achieved by increasing intake of calcium-vitamin D. The active form of vitamin D, the 1, 25-dihydroxy form acts analogously to a steroid hormone and regulates calcium homeostasis. In addition, it acts as a differentiation agent and it inhibited the growth of cancer cells in culture, but in clinical trials hypercalcemia occurred before a therapeutic level (for treatment of leukemia) could be reached. However, various analogs have been synthesized; a trihydroxy vitamin D had differentiating activity without causing hypercalcemia. Further studies of this effect are needed. To complete the vitamin story, studies of the action of vitamin E on breast tissue were reported. In volunteers, vitamin E was given at 1200 IU 5 days before surgery or biopsy of breast tissue. Adipose tissue concentrations of vitamin E (as alpha-tocopherol) were similar in different sites of the body. However, breast tumors with estrogen-negative receptors and poor histological differentiation had lower concentrations of vitamin E than did tumors with positive estrogen receptors and well-differentiated histology.

The posters afforded attendees brief presentations on both experimental and epidemiologic studies of the diet-cancer connection. The international attendance at the conference was reflected in the posters for there were presentations from Canada, France, Mexico and Pakistan. This was in addition to the many posters from research institutions from all parts of the USA.

To summarize, although animal experiments provide sufficient evidence for a role of dietary fat in enhancing mammary carcinogenesis, the situation in humans is not so definitive. However, there are sufficient indications from various epidemiologic studies that higher consumption of fat in the diet is associated with breast cancer incidence, as well as a poorer prognosis in women already diagnosed with this disease. More definitive surveys and trials underway of low-fat diets in large groups of women should resolve the matter.

The Editor

Elizabeth K. Weisburger graduated from Lebanon Valley College, Annville, PA, with a B.S. in Chemistry and received a Ph.D. in Organic Chemistry from the University of Cincinnati. Both institutions subsequently awarded her honorary D.Sc. degrees. After spending two years as a research associate at Cincinnati, she received a postdoctoral fellowship at the National Cancer Institute in Bethesda, MD, where she stayed for almost 40 years. After the fellowship ended, Dr. Weisburger was commissioned in the U.S. Public Health Service, assigned to NCI. At NCI she was involved in studies on metabolism of carcinogens, chemoprevention, and the NCI Bioassay Program, which evolved into the National Toxicology Program to test compounds for possible carcinogenicity. The last eight years of her NCI career were spent in the office of the Director of the Division of Cancer Etiology as an Assistant Director for Chemical Carcinogenesis.

Since retiring from NCI at the beginning of 1989, Dr. Weisburger has worked as a consultant, and she is active in writing and editorial work. She also devotes much of her time to both the local and national American Chemical Society, especially the Division of Chemical Health and Safety, as well as the Threshold Limit Value (TLV) Committee of the American Conference of Governmental Industrial Hygienists.

Dr. Weisburger is a Fellow of the American Association for the Advancement of Science, a member of the American Association for Cancer Research, the American Chemical Society, the American Society for Biochemistry and Molecular Biology, ACGIH, Royal Society of Chemistry, the Society of Toxicology, and smaller groups including Alpha Chi Sigma, Graduate Women in Science, and Iota Sigma Pi.

Dr. Weisburger is listed in American Men and Women of Science, Who's Who of American Women, Who's Who in America, and Women in Chemistry and Physics. Other activities include membership on the Board of Trustees of Lebanon Valley College since 1970 (Board Chair 1985-1989), membership on editorial boards, and some teaching for the National Institutes of Health Graduate School. She is an example of the busy "retired" scientist.

Contents

Chapter 6
Effect of Conjugated Linoleic Acid on Carcinogenesis 59
Joseph A. Scimeca, Henry J. Thompson, and Clement Ip

Chapter 7
**A Possible Mechanism by Which Dietary Fat Can Alter
Tumorigenesis: Lipid Modulation of Macrophage Function** 67
Kent L. Erickson and Neil E. Hubbard

Chapter 8
**Dietary Fatty Acids and Human Breast Cancer Cell
Growth, Invasion, and Metastasis** 83
David P. Rose, Jeanne M. Connolly and Xin-Hua Liu

Chapter 1
Dietary Fat and Breast Cancer: Controversy and Biological Plausibility

DAVID P. ROSE

I. Introduction

The influence which dietary fat may exert on the development of breast cancer continues to provoke heated debate.[1-5] The association between fat consumption and mammary carcinogenesis in experimental animals was first described by Tannenbaum over 50 years ago[6] and has been confirmed repeatedly since then.[7-9] In addition to influencing the promotional stage of carcinogenesis, dietary fat also affects the growth of transplantable rodent mammary carcinomas[10-12] and of human breast cancer cells when injected into athymic nude mice.[13-17]

Despite this experimental support for a relationship between dietary fat intake and breast cancer risk and progression, the results from epidemiological research continue to be conflicting, leading to confusion among health professionals and the public alike. While ecological studies have consistently demonstrated a strong positive correlation between estimates of per capita fat intake and breast cancer risk in different countries,[18-22] and even within the same country,[23] those based on a case-control or cohort design lack consistency.[1,4,24,25]

A related, but distinct, issue is the influence which dietary fat may exert on the progression of an established breast cancer and the prognosis for the postsurgical breast cancer patient. Here the albeit limited epidemiological evidence supports the hypothesis that the level and, perhaps more importantly, the nature of dietary fat consumed do affect human breast cancer cell growth and expression of the metastatic phenotype.[16,17,26-28]

In this overview, we will look more closely at the dietary fat hypothesis of breast cancer risk, examine the controversial aspects of the issue, seek areas of concordance, and discuss some of the support which has been derived from biochemical epidemiology.

David P. Rose ● Division of Nutrition and Endocrinology, American Health Foundation, Valhalla, New York

Diet and Breast Cancer, Edited under the auspices of the
American Institute for Cancer Research, Plenum Press, New York, 1994

II. International, Intranational and Migrant Studies

The earlier international comparisons were mostly of breast cancer mortality rates in different countries and corresponding estimates of dietary fat consumption.[18-20] They showed a strong positive correlation for animal fats, but none for vegetable fat; in our own study,[20] for example, the respective correlation coefficients were 0.76 and 0.18. Despite such encouraging results, these ecological investigations are frequently criticized because of the variable reliability of the mortality data, the lack of precision in the estimates of dietary fat intake, and the potential for confounding by, for example, an unrecognized factor associated with industrialization. Moreover, the finding that the relationship, even if valid, was between breast cancer risk and animal (implicitly saturated) and not vegetable (implicitly unsaturated) fat, consumption appears to be at odds with the results of the animal experiments in which promotion of mammary carcinogenesis is consistently obtained by feeding high levels of omega-6 polyunsaturated fatty acid (linoleic acid)-rich corn oil.[7-9]

Recently, this latter objection appears to have been at least partially resolved by several epidemiological studies which found an association with breast cancer risk for both saturated *and* polyunsaturated fats.[21,22,29] Making the distinction between animal and vegetable fats is, in fact, quite misleading. Under the umbrella term "vegetable fat" are included not only oils rich in mammary tumor-promoting omega-6 fatty acids, but also fish rich in the omega-3 fatty acids which suppress experimental mammary carcinogenesis,[30] and olive oil, a source of omega-9 fatty acid which in some experimental situations appears to enhance mammary tumor development,[31] while in others it has no promotional effect.[32,33]

Japan is a country with one of the lowest breast cancer mortality rates in the world, but this has shown a steady increase since 1960; at the same time, the average fat consumption has increased from 9% of total calories in 1955 to 25% in 1987.[34] Although the increased consumption of red meat has been emphasized in this shift towards a "western" style diet,[23,24] there has also been an increase in the consumption of omega-6 fatty acid-rich vegetable oils. Kamano *et al.*[35] suggested that this may be related to the rising breast cancer risk in Japan. Hirayama[23] examined dietary total fat intake and breast cancer mortality rates in 12 districts in Japan. There was a strong positive correlation (r=0.842), with urban locations exhibiting both the highest age-adjusted mortality rates, and the highest per capita daily fat intakes.

Proliferative forms of benign breast disease have been associated with both an increased risk of breast cancer,[36-38] and frequent consumption of fats,[39, 40] notably those from red meats.[40] Schnitt *et al.*[41] compared the incidence of non-proliferative and proliferative benign breast disease lesions in Japanese women for the years 1974 to 1975, with that existing a decade later. The proliferative forms were significantly more common in the later period, and this was due to an increase principally in women aged less than 40 years. It was suggested that this change may be associated with the "westernization" of the Japanese diet.

Migrant studies provide an instructive exercise for the cancer epidemiologist. When applied to Japanese migrants to Hawaii, they demonstrate a pronounced increase in breast cancer incidence which is evident in the Issei (Japan born) and becomes even more apparent in the Nissei (Hawaii born). Again, these data have been interpreted as reflecting the switch to dietary practices with a western pattern, and specifically an increase in fat consumption.[42]

Similarly, increases in breast cancer risk are seen in migration studies of Chinese,[43] and Polish[44] women and subsequent generations in the United States.

III. Case-Control and Cohort Studies

To the non-epidemiologist, it is, perhaps, somewhat surprising that professionals in this discipline can look at the same series of studies and, while some see diamonds others see coal. Thus, Whittemore and Henderson[4] referred to "several" case-control and cohort studies on the relationship between dietary fat and breast cancer risk, and concluded that, in contrast to the international correlation studies, they provide no support for the postulated association. In a rebuttal, Freedman and colleagues[5] pointed out that in the overview analysis of 12 case-control studies performed by Howe *et al.*,[24] the combined data did show a positive association between fat intake and breast cancer for postmenopausal women (relative risk for the highest versus the lowest quintile, 1.46; p<0.0001). This combined analysis, using the original data records from each study, produced no evidence of a similar association in premenopausal women.

An examination of the conclusions drawn from the individual studies by each group of responsible investigators is instructive. These are available for 9 of the 12 (Table 1); in 3 cases the food records had been collected for a different purpose and had not been analyzed for dietary fat prior to the meta-analysis. In 4 cases[45-48] it was concluded that there was a positive association between fat intake and breast cancer risk. However in one case,[47] in contrast to the combined analysis, the relationship was observed primarily for premenopausal women. A fifth study from Israel[49] showed a complex association with high fat and animal protein intakes interacting with a low fiber intake to produce a positive effect on breast cancer risk. The authors of the remaining 4 reports[50-52] interpreted their results as providing no or little support for the association. One of these from Argentina[51] observed a positive relationship with energy intake, but this effect was distributed across all three macronutrients, and there was a negative association with fiber.

The highly visible Nurses' Health Study[1,53] is producing a plethora of new data on the relationships between diet and chronic diseases. Reports on the cohort study in the context of dietary fat and breast cancer risk have been published at four,[53] and eight,[1] years of follow-up. By the time of the second report there were 1,439 incident cases of breast cancer, with 774 occurring in postmenopausal women. An extremely careful analysis of the data showed no evidence in support of the dietary fat and breast cancer risk hypothesis, even though the same cohort had demonstrated a positive association between total and animal fat and the risk of colon cancer.

We,[54] and others,[3,55] have pointed out that the women in the Nurses' Health Study constitute a homogeneous population with respect to their dietary practices, compared with those represented in the international ecological studies; relatively few consume less than 30% of their calories from fat. We believe, on the basis of experiments with animal models,[56] and studies of dietary fat intake and circulating estrogens,[57] that a reduction in fat intake to 20% of total calories or less is required to favorably influence breast cancer risk. In this sense, then, we have no conflict with the conclusions drawn by Willett *et al.*[1] who acknowledge that their data could not exclude an influence of substantially lower levels of fat consumption, such as below 20% of total energy, on breast cancer risk. They pointed out that for such an effect to hold it is necessary to assume a nonlinear relationship between fat consumption and risk. This is precisely what was found in our animal study utilizing chemically-induced rat mammary carcinomas: tumor development was no different in animals fed 10% or 20% total energy as fat, but was enhanced by a 30% fat-calorie diet with no further stimulation in rats fed a 40% fat-calorie diet.[56]

Jones *et al.*[58] used the National Health and Nutrition Examination Survey I Epidemiologic Follow-up Study cohort of 5,485 women to examine the dietary fat and breast cancer risk hypothesis. On the basis of results obtained with a 24-hour recall method of evaluating fat intake, which may or may not reflect usual intake, they actually found that increased consumption reduced risk. An absence of any association, positive or negative, was reported by Graham *et al.*[59] who used the responses to a mailed questionnaire inquiring on the frequency of consumption of 45 individual food items. Again, it does not seem unreasonable

Table 1. Case-Control Studies Included in the Combined Analysis Performed by Howe *et al.*[24]; Conclusions Drawn From the Individual Studies by the Original Investigators.

Authors of the Original Studies	Location of Study	Interpretation of Study Results in Context of Fat-Breast Cancer Hypothesis
Miller *et al.*[45]	Canada	Causal association
Toniolo *et al.*[46]	Italy	Positive association
Hislop *et al.*[47]	Canada	Association, especially in *premenopausal* women
Rohan *et al.*[48]	Australia	Weak positive association
Lubin *et al.*[49]	Israel	Positive *combined* effects of high fat and animal protein, and low fiber intake
Hirohata *et al.*[50]	Hawaii (Japanese)	No strong support
Hirohata *et al.*[50]	Hawaii (whites)	No strong support
Iscovich *et al.*[51]	Argentina	No specific association
Katsouyanni *et al.*[52]	Greece	No evidence of an association

to speculate on the power of such an instrument to provide a reliable assessment of dietary fat intake.

Three cohort studies of dietary fat and breast cancer have been published from outside the United States. van den Brandt *et al.*[60] concluded that their study in the Netherlands failed to provide any support for dietary fat having a major role in the etiology of postmenopausal breast cancer. However, fat consumption in the Netherlands ranks as one of the highest in the world, and the lowest quintile of participants reported a median intake of 47 g/day (32% of total energy). Even so, a trend towards a positive association for saturated fats and breast cancer risk did, in fact, emerge (p-trend, 0.049).

Data from the Canadian Breast Screening Study, which included a self-administered questionnaire based on 86 food items, were analyzed by Howe *et al.*[25] They observed a positive association between breast cancer and total fat intake that was independent of other energy sources. Subsequently they noted that for postmenopausal women the relative risk estimate for saturated fat, comparing the highest to lowest quintile, was 1.34 (confidence interval: 0.76 – 2.37).[61] This result was quite similar to that obtained from the same authors' combined analysis of 12 case-control studies (1.43; confidence interval, 1.13 – 1.80).[24]

Finland is a country of interest in the context of diet and breast cancer risk, because the traditionally high dietary fiber intake may exert a protective effect in the face of a relatively very high fat consumption.[62] Knekt *et al.*[63] carried out a cohort study of Finnish women aged 20-69 years at entry with a 20 year maximum follow-up. They found a positive, but weak, association between energy-adjusted fat intake and breast cancer risk.

After reviewing the available reports of case-control and cohort studies, we are led to conclude that little of value is likely to emerge from further investigation of the dietary fat and breast cancer hypothesis in populations which are homogeneous with respect to diet and at a relatively high risk for breast cancer. The problem was discussed by Wynder and Stellman,[64] who drew a parallel with the situation which would hold in the case of cigarette smoking and lung cancer risk if the control group included light smokers instead of non-smokers. Their conclusion was that a meaningful study requires that the control group include a sizable proportion of women whose diet is akin to that of the Japanese. Only then would any association between a diet with 30-40% calories as fat and breast cancer become apparent. This deserves careful consideration.

IV. Dietary Fat and Breast Cancer Progression

The effects of dietary fat on breast cancer progression and the prognosis of the post-surgical breast cancer patient are the topics of other articles in this volume.[26,28] They have also been the subject of several other reviews.[65-67] We will note here that the concept that a low-fat dietary intervention may reduce the risk of recurrence after the surgical resection of breast cancer in a postmenopausal patient developed from international comparisons of therapeutic outcome. In 1963, Wynder *et al.* reported that Japanese breast cancer patients had a better survival rate than American patients; all had been treated by modified radical mastectomy, since the study was performed before the advent of systemic adjuvant chemo-hormonal therapy. The differences were evident regardless of the stage of disease at diagnosis. The patients were not stratified on the basis of menopausal status, but when this was done later by Sakamoto *et al.*[69] the difference was found to be limited to postmeno-pausal women.

V. The Search for Biological Plausibility

The acceptance of a hypothesis such as that which proposes an association between dietary fat and breast cancer risk and progression requires the support of a feasible mechanistic explanation. In our particular case, emphasis has been placed upon the relationship between dietary fat and hormones.

It is generally accepted that the estrogenic sex hormones are involved in breast cancer development[57,70,71] and metastasis.[72] Epidemiologic investigations have shown positive relationships between breast cancer risk and early menarche, late menopause, nulliparity, and a delayed first pregnancy,[73] all events which lead to continued exposure of target tissues, such as the breast, to high estrogen levels. In addition, some case-control studies demonstrated differences in the concentration of circulating estrogens,[74,75] and, perhaps more importantly, their biological availability.[75-78]

In contrast to the situation in cities, the diet of Japanese living in rural areas has not undergone the "westernization" referred to earlier, and dietary fat intake remains about 10-20 g/day. Shimizu *et al.*[79] measured serum estrogen concentrations in healthy rural-dwelling postmenopausal Japanese women and postmenopausal white American women. The levels of both serum estradiol and estrone were considerably lower in the Japanese, at low risk for breast cancer, compared with the American women, at a relatively high risk for the disease. Key, Wang, and coworkers,[80,81] made a similar comparison between premenopausal and postmenopausal rural Chinese women and British women of corresponding menopausal status. Again, serum estrogen concentrations were lower in the rural-dwelling Chinese, for whom the estimated dietary fat intake was 15% of total energy; in Britain the average fat intake is approximately 40% of total energy.

Experimental studies to determine directly the influence of dietary factors on blood estrogens are described elsewhere in this volume.[82] Our own studies of both premenopausal,[83] and postmenopausal[84] white American women have indicated that a reduction in dietary fat consumption from a typical 36-40% of total calories to 20% results in significant reductions in serum estradiol levels.

VI. Commentary

It seems unlikely that more is to be gained from case-control or cohort studies sited in countries which have homogeneous populations with respect to their dietary habits.

The next phase of the debate has centered on the place, if any, for a low-fat dietary intervention trial aimed at breast cancer prevention. This is a hotly contested subject,[4,5] and it is the topic of another article in this volume.[85] A frequently expressed criticism of the trial as originally defended by Prentice *et al.*[55] is that it targets postmenopausal women, whereas the likelihood is that if dietary fat does exert an enhancing effect on breast cancer risk, this is likely to operate much earlier in life. However, experience with the antiestrogen tamoxifen indicates that this criticism is not valid. The current trial of tamoxifen as a chemopreventive agent for breast cancer is based, in part, on the observation that adjuvant tamoxifen therapy reduces the risk of a second primary tumor developing in the remaining breast of postsurgically-treated *postmenopausal* breast cancer patients.[86,87] While this strategy is, strictly speaking, one of chemosuppression rather than chemoprevention,[88] it supports the validity of intervention relatively late in life. As we have seen, low-fat dietary intervention, like antiestrogen therapy, results in a reduction in estrogenic bioactivity, which, of itself, is likely to exert a suppressive effect on breast cancer development. Moreover, the putative mechanisms related to eicosanoid biosynthesis, which are described elsewhere in this volume,[28] are also likely to be active late in the process of subclinical tumor progression. This argument does not exclude a role for dietary fat earlier in the neoplastic process, but it only states that later intervention may be effective in arresting the process.

Finally, while we have focussed on fat in this review, it is much more likely that complex interactions between dietary components, both macro- and micronutrients, exist which influence breast cancer risk. Among these, dietary fiber is attracting attention,[49,62] together with fiber-associated phytoestrogens and lignans.[89,90] Also, it may transpire that the most effective approach to breast cancer prevention is one in which dietary manipulation is combined with additive treatment such as the administration of retinoids, antioxidants, and inhibitors of eicosanoid biosynthesis.

References

1. Willett, W.C., D.J. Hunter, M.J. Stampfer, G. Colditz, J.E. Manson, D. Spiegelman, B. Rosner, C.H. Hennekens, and F.E. Speizer, Dietary fat and fiber in relation to risk of breast cancer. An 8-year follow-up, *JAMA* 268:2037 (1992).
2. Howe, G.R., High-fat diets and breast cancer risk. The epidemiologic evidence, *JAMA* 268:2080 (1992).
3. Boyd, N.F., Nutrition and breast cancer, *J. Natl. Cancer Inst.* 85:6 (1993).
4. Whittemore, A.S. and B.E. Henderson, Dietary fat and breast cancer: where are we?, *J. Natl. Cancer Inst.* 85:762 (1993).
5. Freedman, L.S., R.L. Prentice, C. Clifford, W. Harlan, M. Henderson, and J. Rossouw, Dietary fat and breast cancer: where we are, *J. Natl. Cancer Inst.* 85:764 (1993).
6. Tannenbaum, A., The genesis and growth of tumors. III. Effects of a high fat diet, *Cancer Res.* 2:468 (1942).
7. Cohen, L.A., Dietary fat and mammary cancer, *in: Diet, Nutrition and Cancer: A Critical Evaluation, Vol. 1 Macronutrients and Cancer*, B.S. Reddy and L.A. Cohen, eds., CRC Press, Boca Raton (1986).
8. Welsch, C.W., Dietary fat, calories, and mammary gland tumorigenesis, *in: Exercise, Calories, Fat, and Cancer*, M.M. Jacobs, ed., Plenum Press, New York (1992).
9. Cave, W.T. Jr., Dietary fat effects on animal models of breast cancer, *in: Diet and Breast Cancer*, E.K. Weisburger, ed., Plenum Press, New York (1994).

10. Rao, G.A. and S. Abraham, Enhanced growth rate of transplanted mammary adenocarcinoma induced in C3H mice by dietary linoleate, *J. Natl. Cancer Inst.* 56:431 (1976).
11. Hillyard, L.A. and S. Abraham, Effect of dietary polyunsaturated fatty acids on growth of mammary adenocarcinomas in mice and rats, *Cancer Res.* 39:4430 (1979).
12. Karmali, R.A., J. Marsh, and C. Fuchs, Effect of omega-3 fatty acids on growth of a rat mammary tumor, *J. Natl. Cancer Inst.* 73:457 (1984).
13. Pritchard, G.A., D.L. Jones, and R.E. Mansel, Lipids in breast cancer, *Br. J. Surg.* 76:1069 (1989).
14. Borgeson, C.E., L. Pardini, R.S. Pardini, and R.C. Reitz, Effects of dietary fish oil on human mammary carcinoma and on lipid-metabolizing enzymes, *Lipids* 24:290 (1989).
15. Gonzalez, M.J., R.A. Schemmel, J.I. Gray, L. Dugan Jr., L.G. Sheffield, and C.W. Welsch, Effect of dietary fat on growth of MCF-7 and MDA-MB-231 human breast carcinomas in athymic nude mice: relationship between carcinoma growth and lipid peroxidation product levels, *Carcinogenesis* 12:1231 (1991).
16. Rose, D.P., J.M. Connolly, and C.L. Meschter, Effect of dietary fat on human breast cancer growth and lung metastasis in nude mice, *J. Natl. Cancer Inst.* 83:1491 (1991).
17. Rose, D.P., M.A. Hatala, J.M. Connolly, and J. Rayburn, Effects of diets containing different levels of linoleic acid on human breast cancer growth and lung metastasis in nude mice, *Cancer Res.* 53:4686 (1993).
18. Armstrong, B. and R. Doll, Environmental factors and cancer incidence and mortality in different countries, with special reference to dietary practices, *Int. J. Cancer* 15:617 (1975).
19. Hems, G., The contributions of diet and childbearing to breast cancer rates, *Br. J. Cancer* 37:974 (1978).
20. Rose, D.P., A.P. Boyar, and E.L. Wynder, International comparisons of mortality rates for cancer of the breast, ovary, prostate, and colon, and per capita food consumption, *Cancer* 58:2363 (1986).
21. Prentice, R.L. and L. Sheppard, Dietary fat and cancer: consistency of the epidemiologic data, and disease prevention that may follow from a practical reduction in fat consumption, *Cancer Causes Control* 1:81 (1990).
22. Sasaki, S., M. Horacsek, and H. Kesteloot, An ecological study of the relationship between dietary fat intake and breast cancer mortality, *Prev. Med.* 22:187 (1993).
23. Hirayama, T., Epidemiology of breast cancer with special reference to the role of diet, *Prev. Med.* 7:173 (1978).
24. Howe, G.R., T. Hirohata, T.G. Hislop, J.M. Iscovich, J.-M. Yuan, K. Katsouyanni, F. Lubin, E. Marubini, B. Modan, T. Rohan, P. Toniolo, and Y. Shunzhang, Dietary factors and risk of breast cancer: Combined analysis of 12 case-control studies, *J. Natl. Cancer Inst.* 82:561 (1990).
25. Howe, G.R., C.M. Friedenreich, M. Jain, and A.B. Miller, A cohort study of fat intake and risk of breast cancer, *J. Natl. Cancer Inst.* 83:336 (1991).
26. Chlebowski, R.T., Dietary fat intake reduction for patients with resected breast cancer, *in*: *Diet and Breast Cancer*, E.K. Weisburger, ed., Plenum Press, New York (1994).
27. Rose, D.P. and J.M. Connolly, Effects of dietary omega-3 fatty acids on human breast cancer growth and metastases in nude mice, *J. Natl. Cancer Inst.* 85:1743 (1993).
28. Rose, D.P., J.M. Connolly, and X.-H. Liu, Dietary fatty acids and human breast cancer cell growth, invasion, and metastasis, *in*: *Diet and Breast Cancer*, E.K. Weisburger, ed., Plenum Press, New York (1994).
29. Hursting, S.D., M. Thornquist, and M.M. Henderson, Types of diet fat and the incidence of cancer at five sites, *Prev. Med.* 19:242 (1990).
30. Jurkowski, J.J. and W.T. Cave Jr., Dietary effects of menhaden oil on the growth and membrane lipid composition of rat mammary tumors, *J. Natl. Cancer Inst.* 74:1145 (1985).
31. Carroll, K.K. and H.T. Khor, Effects of level and type of dietary fat on incidence of mammary tumors induced in female Sprague-Dawley rats by 7,12-dimethylbenz[a]anthracene, *Lipids* 6:415 (1971).

32. Cohen, L.A., D.O. Thompson, Y. Maeura, K. Choi, M.E. Blank, and D.P. Rose, Dietary fat and mammary cancer. I. Promoting effects of different dietary fats on N-nitrosomethylurea-induced rat mammary tumorigenesis, *J. Natl. Cancer Inst.* 77:33 (1986).
33. Tinsley, I.J., J.A. Schmitz, and D.A. Pierce, Influence of dietary fatty acids on the incidence of mammary tumors in the C3H mouse, *Cancer Res.* 41:1460 (1981).
34. Wynder, E.L., Y. Fujita, R.E. Harris, T. Hirayama, and T. Hiyama, Comparative epidemiology of cancer between the United States and Japan. A second look, *Cancer* 67:746 (1991).
35. Kamano, K., H. Okuyama, R. Konishi, and H. Nagasawa, Effects of a high-linoleate and a high α-linolenate diet on spontaneous mammary tumorigenesis in mice, *Anticancer Res.* 9:1903 (1989).
36. Dupont, W.D. and D.L. Page, Risk factors for breast cancer in women with proliferative breast disease, *N. Engl. J. Med.* 312:146 (1985).
37. Webber, W. and N. Boyd, A critique of the methodology of studies of benign breast disease and breast cancer risk, *J. Natl. Cancer Inst.* 77:397 (1986).
38. Carter, C.L., D.K. Carle, M.S. Micozzi, A. Schatzkin, and P.R. Taylor, A prospective study of the development of breast cancer in 16,692 women with benign breast disease, *Am. J. Epidemiol.* 128:467 (1988).
39. Lubin, F., Y. Wax, E. Ron, M. Black, A. Chetrit, N. Rosen, E. Alfandary, and B. Modan, Nutritional factors associated with benign breast disease etiology: a case-control study. *Am. J. Clin. Nutr.* 50:551 (1989).
40. Hislop, T.G., P.R. Band, M. Deschamps, V. Ng, A.J. Coldman, A.J. Worth, and T. Labo, Diet and histologic types of benign breast disease defined by subsequent risk of breast cancer, *Am. J. Epidemiol.* 131:263 (1990).
41. Schnitt, S.J., A. Jimi, and M. Kojiro, The increasing prevalence of benign proliferative breast lesions in Japanese women, *Cancer* 71:2528 (1993).
42. Wynder, E.L., D.P. Rose, and L.A. Cohen, Diet and breast cancer in causation and therapy, *Cancer* 58:1804 (1986).
43. King, H. and F.B. Locke, Cancer mortality among Chinese in the United States, *J. Natl. Cancer Inst.* 65:1141 (1980).
44. Staszewski, J. and W. Haenszel, Cancer mortality among Polish born in the United States, *J. Natl. Cancer Inst.* 35:292 (1965).
45. Miller, A.B., A. Kelly, and N.W. Choi, A study of diet and breast cancer, *Am. J. Epidemiol.* 107:499 (1978).
46. Toniolo, P., E. Riboli, F. Protta, M. Charrel, and A.P.M. Cappa, Calorie-providing nutrients and risk of breast cancer, *J. Natl. Cancer Inst.* 81:278 (1989).
47. Hislop, T.G., A.J. Coldman, and J.M. Elwood, Childhood and recent eating patterns and risk of breast cancer, *Cancer Detect. Prev.* 9:47 (1986).
48. Rohan, T.E., A.J. McMichael, and P.A. Baghurst, A population-based case-control study of diet and breast cancer in Australia, *Am. J. Epidemiol.* 128:478 (1988).
49. Lubin, F., Y.K. Wax, and B. Modan, Role of fat, animal protein and dietary fiber in breast cancer etiology: A case-control study, *J. Natl. Cancer Inst.* 77:605 (1986).
50. Hirohata, T., A. Nomura, J.H. Hankin, L.N. Kolonel, and J. Lee, An epidemiologic study on the association between diet and breast cancer, *J. Natl. Cancer Inst.* 78:595 (1987).
51. Iscovich, J., G.R. Howe, and J.M. Kaldor, A case-control study of breast cancer in Argentina, *Int. J. Cancer* 44:770 (1989).
52. Katsouyanni, K., D. Trichopoulos, P. Boyle, E. Xirouchaki, A. Trichopoulou, B. Lisseos, S. Vasilaros, and B. MacMahon, Diet and breast cancer: a case-control study in Greece, *Int. J. Cancer* 38:815 (1986).
53. Willett, W.C., M.J. Stampfer, G.A. Colditz, B.A. Rosner, C.H. Hennekens, and F.E. Speizer, Dietary fat and the risk of breast cancer, *N. Engl. J. Med.* 316:22 (1987).
54. Wynder, E.L., E. Taioli, and D.P. Rose, Breast cancer—the optimal diet, *in: Exercise, Calories, Fat, and Cancer*, M.M. Jacobs, ed., Plenum Press, New York (1992).
55. Prentice, R.L., F. Kakar, S. Hursting, L. Sheppard, R. Klein, and L.H. Kushi, Aspects of the rationale for the Women's Health Trial, *J. Natl. Cancer Inst.* 80:802 (1988).

56. Cohen, L.A., K. Choi, J.H. Weisburger, and D.P. Rose, Effect of varying proportions of dietary fat on the development of *N*-nitrosomethylurea-induced rat mammary tumors, *Anticancer Res.* 6:215 (1986).
57. Rose, D.P., Diet, hormones, and cancer, *Annu. Rev. Public Health* 14:1 (1993).
58. Jones, D.Y., A. Schatzkin, S.B. Green, G. Block, L.A. Brinton, R.G. Ziegler, R. Hoover, and P.R. Taylor, Dietary fat and breast cancer in the National Health and Nutrition Examination Survey. I. Epidemiologic Follow-up Study, *J. Natl. Cancer Inst.* 79:465 (1987).
59. Graham, S., M. Zielezny, J. Marshall, R. Priore, J. Freudenheim, J. Brasure, B. Haughey, P. Nasca, and M. Zdeb, Diet in the epidemiology of postmenopausal breast cancer in the New York State cohort, *Am. J. Epidemiol.* 136:1327 (1992).
60. van den Brandt, P.A., P. van't Veer, R.A. Goldbohm, E. Dorant, A. Volovics, R.J.J. Hermus, and F. Sturmans, A prospective cohort study on dietary fat and the risk of postmenopausal breast cancer, *Cancer Res.* 53:75 (1993).
61. Howe, G.R., Dietary fat and cancer: Response, *J. Natl. Cancer Inst.* 83:1035 (1991).
62. Rose, D.P., Dietary fiber and breast cancer, *Nutr. Cancer* 13:1 (1990).
63. Knekt, P., D. Albanes, R. Seppänen, A. Aromaa, R. Järviven, L. Hyvönen, L. Teppo, and E. Pukkala, Dietary fat and risk of breast cancer, *Am. J. Clin. Nutr.* 52:903 (1990).
64. Wynder, E.L. and S.D. Stellman, The "over-exposed" control group, *Am. J. Epidemiol.* 135:459 (1992).
65. Wynder, E.L. and L.A. Cohen, A rationale for dietary intervention in the treatment of postmenopausal breast cancer patients, *Nutr. Cancer* 3:195 (1982).
66. Chlebowski, R.T., D.P. Rose, I.M. Buzzard, G.L. Blackburn, W. Insull Jr., M. Grosvenor, R. Elashoff, and E.L. Wynder, Adjuvant dietary fat intake reduction in postmenopausal breast cancer patient management, *Breast Cancer Res. Treat.* 20:73 (1992).
67. Cohen, L.A., D.P. Rose, and E.L. Wynder, A rationale for dietary intervention in postmenopausal breast cancer patients: An update, *Nutr. Cancer* 19:1 (1993).
68. Wynder, E.L., T. Kajitani, J. Kuno, J.C. Lucas, and A. DePalo, A comparison of survival rates between American and Japanese patients with breast cancer, *Surg. Gynecol. Obstet.* 117:196 (1963).
69. Sakamoto, G., H. Sugano, and W.H. Hartman, Comparative clinicopathological study of breast cancer among Japanese and American females, *Jpn. J. Cancer* 25:161 (1979).
70. Dickson, R.B. and M.E. Lippman, Estrogenic regulation of growth and polypeptide growth factor secretion in human breast carcinoma, *Endocr. Rev.* 8:29 (1987).
71. Henderson, B.E., R. Ross, and L. Bernstein, Estrogens as a cause of human cancer: The Richard and Hilda Rosenthal Foundation Award Lecture, *Cancer Res.* 48:246 (1988).
72. Litherland, S. and M. Jackson, Antiestrogens in the management of hormone-dependent breast cancer, *Cancer Treat. Rev.* 15:183 (1987).
73. Kelsey, J.L. and G.S. Berkowitz, Breast cancer epidemiology, *Cancer Res.* 48:5615 (1988).
74. Adami, H.-O., E.D.B. Johansson, J. Vegelius, and A. Victor, Serum concentrations of estrone, androstenedione, testosterone and sex-hormone-binding globulin in postmenopausal women with breast cancer and in age-matched controls, *Ups. J. Med. Sci.* 84:259 (1979).
75. Moore, J.W., G.M.G. Clark, R.D. Bulbrook, J.L. Hayward, J.T. Murai, G.L. Hammond, and P.K. Siiteri, Serum concentrations of total and non-protein-bound oestradiol in patients with breast cancer and in normal controls, *Int. J. Cancer* 29:17 (1982).
76. Jones, L.A., D.M. Ota, G.A. Jackson, P.M. Jackson, K. Kemp, D.E. Anderson, S.K. McCamant, and D.H. Bauman, Bioavailability of estradiol as a marker for breast cancer risk assessment, *Cancer Res.* 47:5224 (1987).
77. Ota, D.M., L.A. Jones, G.L. Jackson, P.M. Jackson, K. Kemp, and D. Bauman, Obesity, non-protein-bound estradiol levels, and distribution of estradiol in the sera of breast cancer patients, *Cancer* 57:558 (1986).
78. Takatani, O., H. Kosano, T. Okumoto, K. Akamatsu, S. Tamakuma, and H. Hiraide, Distribution of estradiol and percentage of free testosterone in sera of Japanese women: preoperative breast cancer patients and normal controls, *J. Natl. Cancer Inst.* 79:1199 (1987).

79. Shimizu, H., R.K. Ross, L. Bernstein, M.C. Pike, and B.E. Henderson, Serum oestrogen levels in postmenopausal women: comparison of American whites and Japanese in Japan, *Br. J. Cancer* 62:451 (1990).

80. Key, T.J.A., J. Chen, D.Y. Wang, M.C. Pike, and J. Boreham, Sex hormones in women in rural China and in Britain, *Br. J. Cancer* 62:631 (1990).

81. Wang, D.Y., T.J.A. Key, M.C. Pike, J. Boreham, and J. Chen, Serum hormone levels in British and rural Chinese females, *Breast Cancer Res. Treat.* 18:S41 (1991).

82. Goldin, B.R. and S.L. Gorbach, Hormone studies and the diet and breast cancer connection, *in: Diet and Breast Cancer*, E.K. Weisburger, ed., Plenum Press, New York (1994).

83. Rose, D.P., A.P. Boyar, C. Cohen, and L.E. Strong, Effect of a low-fat diet on hormone levels in women with cystic breast disease. I. Serum steroids and gonadotropins, *J. Natl. Cancer Inst.* 78:623 (1987).

84. Rose, D.P., J.M. Connolly, R.T. Chlebowski, I.M. Buzzard, and E.L. Wynder, The effects of a low-fat dietary intervention and tamoxifen adjuvant therapy on the serum estrogen and sex hormone-binding globulin concentrations of postmenopausal breast cancer patients, *Breast Cancer Res. Treat.* 27:253 (1993).

85. Prentice, R.L., Dietary fat reduction as a hypothesis for the prevention of postmenopausal breast cancer, and a discussion of hypothesis testing research strategies, *in: Diet and Breast Cancer*, E.K. Weisburger, ed., Plenum Press, New York (1994).

86. Early Breast Cancer Trialists' Collaborative Group, Systemic treatment of early breast cancer by hormonal, cytotoxic, or immune therapy: 133 randomised trials involving 31,000 recurrences and 24,000 deaths among 75,000 women, *Lancet* 339:71 (1992).

87. Nayfield, S.G., J.E. Korp, L.G. Ford, F.A. Dorr, and B.S. Kramer, Potential role of tamoxifen in prevention of breast cancer, *J. Natl. Cancer Inst.* 83:1450 (1991).

88. Jordan, V.C., Chemosuppression of breast cancer with long-term tamoxifen therapy, *Prev. Med.* 20:3 (1991).

89. Adlercreutz, H., H. Honjo, A. Higashi, T. Fotsis, E. Hämäläinen, T. Hasegawa, and H. Okada, Urinary excretion of lignans and isoflavonoid phytoestrogens in Japanese men and women consuming a traditional Japanese diet, *Am. J. Clin. Nutr.* 54:1093 (1991).

90. Rose, D.P., Dietary fiber, phytoestrogens, and breast cancer, *Nutrition* 8:47 (1992).

Chapter 2
Dietary Fat Intake Reduction for Patients with Resected Breast Cancer

ROWAN T. CHLEBOWSKI

I. Introduction

In early 1994, a clinical trial has been initiated to definitively address, for the first time, whether a dietary fat intake reduction program can influence the clinical course of patients with breast cancer. This study, the Women's Intervention Nutrition Study (WINS) is based on a hypothesis which is directly supported by laboratory, animal and epidemiological observations.[1-3] In addition, the feasibility of prospectively testing this hypothesis has been established by a series of pilot clinical studies.[2,4-8] These results represent a culmination of over a decade of research activity and are outlined in more detail below.

II. Dietary Fat Intake and Breast Cancer Patient Outcome

A. Epidemiological Evidence

The original impetus to consider reduction in dietary fat intake as a potential therapy for established breast cancer arose from observations comparing stage by stage outcome for patients with resected breast cancer in countries with high dietary fat intakes, such as Western Europe and the United States, to countries where dietary fat intakes were low such as Japan.[9-13] Several groups of investigators found a relatively large and consistent reduction in relapse rate and improvement in overall survival, especially favoring postmenopausal populations of breast cancer patients diagnosed and followed in countries with lower dietary fat intakes.[9-13] The idea that environmental rather than genetic factors were responsible for this difference has been supported by observations in women migrating from countries with differing dietary fat intakes.[14] In these situations, the favorable prognosis of women from countries with lower dietary intakes appears not to be maintained when dietary patterns consistent with higher fat intakes are adopted.[15]

Rowan T. Chlebowski ● Division of Medical Oncology, Harbor-UCLA Medical Center, Torrance, California

Diet and Breast Cancer, Edited under the auspices of the
American Institute for Cancer Research, Plenum Press, New York, 1994

Obviously, many other factors besides diet may play a role in differing outcomes seen among various countries. However, many potential confounding factors such as differences in therapy or tumor histology, as explanations of the observed difference have been considered and rejected.[16] In addition the magnitude of the differences in outcome between countries with low and high dietary fat intake is quite substantial. A life table representation of results from the most recent report in this area,[12] comparing survival of Japanese (low dietary fat) versus United Kingdom (high dietary fat) breast cancer patients diagnosed in a localized stage is illustrated in Figure 1. For this population, largely composed of stage II breast cancer patients, 10 year survival in Japanese patients was 73% compared to a 10 year survival of 52% for patients from the United Kingdom. To provide an estimate of the magnitude of this difference in contrast to currently available therapies, the survival for patients with resected breast cancer randomized to receive adjuvant tamoxifen treatment (a "standard" intervention) is compared to the survival of patients randomized to control/ placebo in Figure 1. These latter results have been abstracted from the Early Breast Cancer Trialists' Collaborative Group report summarizing an over 16,000 patient experience in randomized trials.[12] As seen, survival for the control patients in the UK and the Early Breast Cancer Trialists' experience was nearly identical (53 vs 52%). While tamoxifen resulted in a significant survival benefit compared to no adjuvant therapy (p<0.0001), the absolute difference was only 6% (59% vs 53% of patients alive at 10 years by life table analysis). These results can be compared to the more than three-fold greater difference in absolute survival of 21% (73% vs 52%) seen in the country by country comparison of the Japanese and UK outcome. Obviously, these data sets are derived by two different methodologies and cannot be compared directly. However, they are useful in indicating the magnitude of the potential influence on patient outcome which might be achieved if the dietary hypothesis linking increased fat intake with adverse clinical outcome can be supported.

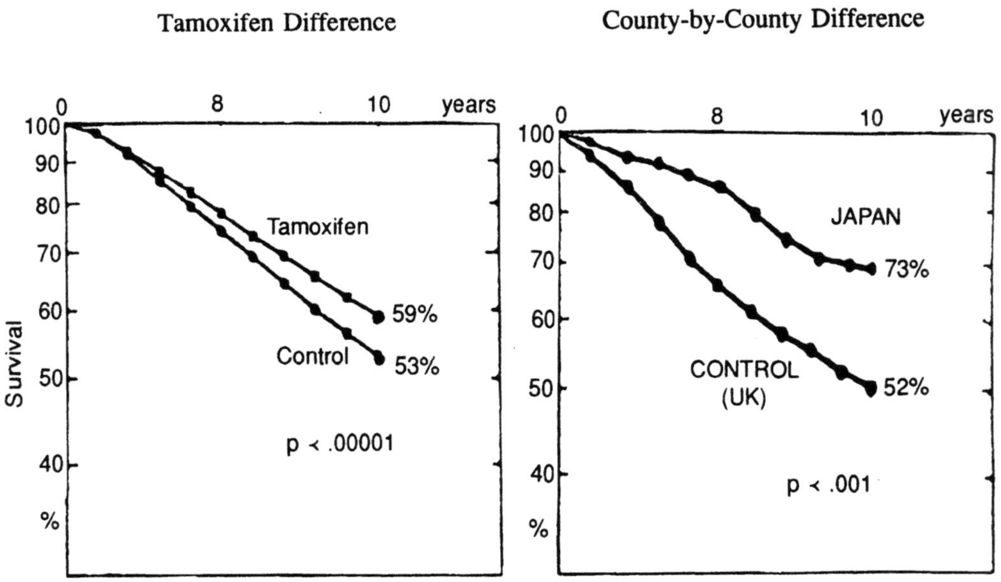

Data summarized from references 12 and 17.

Figure 1. Comparing the magnitude of tamoxifen effect to country-by-country differences in stage-matched breast cancer patient outcome.

At this point it is worth emphasizing that the consideration of dietary fat intake reduction as a breast cancer preventive approach and dietary fat intake reduction as a breast cancer therapeutic approach represent two distinct hypotheses which require separate bodies of evidence for support.[1,3,18,19] The scope of the current discussion is limited to the role of dietary fat intake on the growth and metastatic spread of established breast cancer. Thus, the information presented in this report relates to reduction in dietary fat intake as a potential breast cancer therapy rather than as a breast cancer preventive approach.

B. Laboratory and Animal Experimental Evidence

In this regard, a considerable body of laboratory and experimental evidence has been developed over the past years to support a link between dietary fat intake and breast cancer growth and metastatic spread.[20-22] In animal systems, the most consistent component of fat intake linked to breast cancer growth has been the polyunsaturated fatty acids, with most attention and the greatest weight of evidence pointing to linoleic acid as a potentially unique stimulant of mammary cancer growth in animals.[20] In a representative series of experiments, Rose and colleagues[21] demonstrated that linoleic acid can stimulate directly growth of a human breast cancer derived cell line *in vitro*. Additionally, their research group has demonstrated that a human-derived breast cancer cell line more commonly metastasizes when placed in nude mice given diets high in linoleic acid compared to animals on reduced linoleic acid intakes.[22] Work by other investigators is generally consistent with a role for unsaturated fatty acids in stimulating growth of mammary cancers in animal systems when compared to saturated or monounsaturated fatty acids.[21,22] The animal evidence has been extensively reviewed.[20] However, the limited clinical or epidemiological studies in women largely preclude extrapolation regarding which categories of fat may play a role in influencing breast cancer metastases and growth in women.

In terms of biological plausibility, a series of potential mediators of the effect reduction in dietary fat intake has on breast cancer growth has been outlined. These mechanisms include changes in circulatory hormonal levels,[23-29] changes in cell membrane structure and function (both for breast cancer cells themselves as well as host immune modulating cells), changes in prostaglandin and cytokine/monokine levels and altered regulatory gene expression.[1,3,20] Thus, reasonable mechanisms through which dietary fat intake may influence breast cancer growth have experimental support.

C. Clinical Evidence

For physicians, the most accessible data supporting a relationship between dietary fat reduction and breast cancer patient outcome come from a series of clinical observations (Table 1). These reports relate breast cancer patient outcome to: geographic areas associ-

Table 1. Clinical Data Supporting a Relationship Between Dietary Fat Reduction and Breast Cancer Patient Outcome

Relationship of patient outcome and:

1. Geographic areas associated with low versus high dietary fat intake

2. Obesity at the time of resection

3. Weight gain following resection

4. Dietary fat intake at resection (seen in some reports).

ated with low versus high dietary fat intake, as reviewed;[9-13] obesity at the time of resection;[30,31] weight gain following resection;[32-37] and dietary fat intake at the time of resection, seen in some reports.[15,38-40]

D. Obesity Associated with Breast Cancer Patient Outcome

Among examined factors, increased body weight and/or obesity has been consistently associated with increased risk for breast cancer recurrence.[30] Recently, a 10 year follow-up of 928 breast cancer patients treated at the Sloan-Kettering Cancer Center identified that obesity (>25% above the optimum weight) resulted in identification of a population with a 25% greater stage-for-stage risk of breast cancer recurrence and death. A study of similar design involving over 8,000 women with breast cancer also identified a death rate nearly two times higher for women with initially localized cancer if they were in the highest percentile of bone mass compared to those in the lowest percentile.[41] Most recently, this effect of obesity as an adverse prognostic factor for patients with breast cancer was observed in a large population of breast cancer patients at M.D. Anderson Cancer Center, even in patients receiving contemporary adjuvant chemotherapy.[31] Such large, recently reported studies are consistent with the majority of reports, which identify an adverse prognostic role for obesity in relationship to breast cancer patient outcome. Whether reduction in body weight after diagnosis might reduce the risk of subsequent recurrence and improve clinical outcome remains an unanswered question.

E. Weight Gain Associated with Systemic Adjuvant Therapy and Patient Outcome

Weight gain of substantial magnitude is commonly associated with systemic adjuvant therapy use, especially CMF-like combinations.[32-37] After early reports[33] suggested an association with patient outcome, Camoriano and colleagues[32] reported decreased survival for breast cancer patients gaining more than the group median amount of weight during their adjuvant chemotherapy.

Given the difficulty in accessing caloric intake accurately using currently available dietary assessment techniques, the mechanisms underlying the weight increase associated with breast cancer resection and systemic adjuvant therapy use have not been clearly defined.[3,6] Potential mechanisms include increased caloric intake, change in physical activity, and change in host metabolic status, as reviewed.[36] Regardless of the etiology, data from two feasibility studies have indicated that relatively intensive dietary intervention programs targeted to reduction in dietary fat intake can effect significant weight reduction and abrogate the weight increase associated with systemic adjuvant therapy.[2,4,7] The definite outcome studies now beginning, with patient entry targeted to dietary fat intake for patients with localized breast cancer will determine whether such weight reduction is associated with a beneficial clinical outcome as well.

F. Dietary Fat Intake at Time of Resection and Breast Cancer Patient Outcome

Despite the strength of the laboratory and initial epidemiologic evidence, as reviewed, relating a role for dietary fat intake to breast cancer outcome, the evidence from one area, prospective cohort studies, has been less consistent. The difficulties in conducting such studies have been well documented[42] and they include: methodologic difficulty in accurately measuring dietary intake (frequently using less than ideal assessment methods), small sample size of study populations, usual collection of dietary data at only one time following diagnosis, and most importantly, the relatively limited range of dietary fat intake found in populations identified in a single geographical area. Given these methodologic concerns, several studies have nonetheless suggested that higher dietary fat intakes were inversely related to prognosis.[15,38-40] Other studies utilizing similar study designs have found no clear relationship between fat consumption and prognosis.[43-46] The issue of the limited range of dietary intakes identified in such reports is a critical one since in most of these prospective cohort studies, the lowest group for dietary fat intake frequently exceeds the control dietary intake found in prospective intervention studies.[2,4] Thus, if a threshold for an effect of reduction in dietary fat intake on established breast cancer growth is present, it is unlikely that such prospective cohort studies will be able to address this issue definitively.

III. Mechanisms Potentially Mediating Dietary Fat Effects on Breast Cancer

Potential mechanisms by which dietary fat intake reduction could mediate changes in breast cancer growth and metastatic spread include alterations in hormone levels, changes in membrane structure and function, changes in cytokine and monokine activity along with prostaglandins, and changes in regulatory gene expression.[1,3,20] With respect to hormonal level alteration, estrogen levels are low in countries with low-fat intakes and higher in areas with higher dietary fat intakes.[23,29] However, if estrogen alteration is the only critical variable influenced by a low-fat diet, then a reasonable argument could be to treat patients with the "anti-estrogen" tamoxifen rather than attempting to implement a behavioral change program involving reduction in dietary fat intake. However, it is well recognized that the mechanisms of tamoxifen action are quite complex,[47] and in many areas (such as its effect on lipid profile and bone marrow density) tamoxifen acts as an estrogen.[47,48] Nonetheless, the best clinical evidence suggesting that dietary fat intake and tamoxifen do not share a single mechanism influencing breast cancer growth comes from a large tamoxifen study in Japan, The ACETBC Japan Trial. This trial, with primary data presented in the Overview Analysis,[17] represents the largest randomized study comparing tamoxifen to placebo for patients with early stage resected breast cancer. In this trial, an extremely large tamoxifen effect (Table 2) was seen (approximately twice as great a tamoxifen effect when compared to other studies evaluating tamoxifen in different countries). This substantial effect was seen in the face of the low dietary fat intake found in a Japanese population.[14] Thus, if tamoxifen and dietary fat shared the same mechanism of action on breast cancer, little effect of tamoxifen in the face of a low-fat diet would be expected. In fact, a greater than expected effect was seen suggesting strongly that, on clinical grounds, tamoxifen and dietary fat intake reduction may influence breast cancer growth through separate mechanisms.

Table 2. Effectiveness of Tamoxifen as Adjuvant Therapy in ACETBC (Japan) Trial (Low Dietary Fat Intake Population) Compared to Tamoxifen Trials in Other Countries

Population Study	Number of Patients Randomized	Reduction in Annual Odds of Death
ACETBC (Japan)	3,436	38%
All other (19 total)	12,037	16%

Substantial tamoxifen effect was seen in the context of the low dietary fat intake found in the Japanese population.

Data from Reference 17

IV.   Dietary Fat Intake Reduction in Women With Breast Cancer: Feasibility

Against this background information, the question remains as to whether a free-living population with resected breast cancer will be able to adhere sufficiently to a dietary program of reduction in fat intake to allow the hypothesis relating dietary fat to breast cancer patient outcome to be tested. This issue has recently been addressed in feasibility studies originating in both the United States and Europe.[2,4,7] In addition,[49-52] adherence of women without a cancer diagnosis to other trials involving reduction in dietary fat intake helps to determine the feasibility of such an intervention.[53,54]

For any dietary modification trial, it must be explicitly stated that what is being tested is a hypothesis related to changes in eating patterns rather than directly testing hypotheses related to individual nutrients. This is the case since a change in dietary fat intake results in a cascade of changes in intake of other nutrients.[2,49,52,54] However, this fact does not lessen the clinical significance of identifying a potential significance of lower dietary fat intake on breast cancer patient outcome.

V.   The Women's Intervention Nutrition Study (WINS)

The study outline for the trial addressing the feasibility of implementing a program on reduction of dietary fat intake in a population of postmenopausal women with resected breast cancer which was conducted in the United states is outlined in Figure 2. In this Women's Intervention Nutrition Study (WINS), a population of postmenopausal women with localized

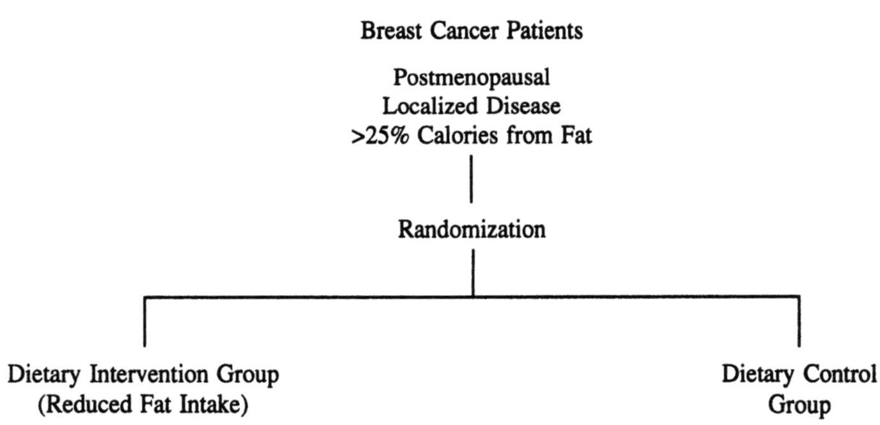

Dietary Fat Intake as Study Endpoint

Conventional Systemic Adjuvant Therapy Given

Sample Size: 2,000 women with resected breast cancer

Objective: To test whether dietary fat intake reduction will reduce breast cancer recurrence and increase patient survival

Dietary Goal: Reduction to 15% of calories as fat

Figure 2. WINS definitive outcome study design.

and resected breast cancer (stage I-IIIa) who had received total mastectomy or lumpectomy plus radiation therapy with axillary dissection were eligible for the study. Additional eligibility factors necessitated by the dietary nature of the intervention included: accessible geographically and by telephone for a two year period; medically able to accept either dietary group assignment; able to provide dietary data as determined by an ability to provide a 4-day food record prior to entry; and a baseline dietary intake of at least 25% of calories from fat determined by a 4-day food record prior to entry. The dietary fat intake target for the dietary intervention group was 20% of calories from fat for the initial 180 patients entered on the feasibility trial and 15% of calories from fat for the subsequent 110 patients. This dietary intervention has been previously described; essentially it involved implementing a "low-fat eating plan" with four bi-weekly one-to-one visits with a nutritionist (Registered Dietician (RD) equivalent).[2,4,49]

The population was unselected, given the random nature of the dietary programs (patients could be randomized to either Intervention or Control Dietary Groups) and the fact that no "run-in" period to determine early adherence to the dietary program was involved in the protocol design.

The total cost of this intervention as delivered in the trial which required only four Registered Dietician contact hours was approximately $400. This amount can be compared to the baseline costs of management for a patient with localized and resectable breast cancer which include: initial surgery, radiation therapy, systemic chemotherapy and continuous tamoxifen treatment. Such a combined modality approach is frequently indicated, resulting in costs which can be estimated at $25,000.[55] Thus, such a program represents an extremely modest 1.5% increase in the incremental cost of managing typical patients with resected breast cancer.

The results of this WINS feasibility effort have been recently published[2] and they are summarized in Table 3. Two hundred and ninety patients were randomized from eight participating institutions to either a Dietary Intervention Group (fat intake reduction) or a Dietary Control Group (counseled for nutritional adequacy only). A reduction in dietary fat intake was rapidly achieved in the intervention group patients (fat consumption decreased from 32% of calories at baseline to 20% of calories at 3 months, p<0.05). This level of dietary fat intake reduction was maintained throughout the 24 month observation period (Table 3). As seen, dietary fat intake remained relatively unchanged in the Control Dietary Group during this period. In addition to total dietary fat, the fat reduction program influenced intake of other nutrients as well. All categories of fat (saturated, polyunsaturated

Table 3. Change in Dietary Fat Intake and Weight After 24 Months in the Women's Intervention Nutrition Study

Parameters	Dietary Fat Reduction		Dietary Control	
	Baseline	24 Mos	Baseline	24 Mos
% Calories from fat	35.0 ± 6.7	24.7 ± 6.0*	34.1 ± 5.9	31.3 ± 8.5
Fat grams	68.7 ± 24.5	31.5 ± 9.1*	68.7 ± 25.1	58.1 ± 28.1
Body weight	70.0 ± 13.9	67.7 ± 14.0	68.7 ± 11.9	69.4 ± 13.2

* Changes are significantly different between two groups (p<0.05).

Data summarized from Reference 2.

including linoleic acid, and monounsaturated) were reduced 50% or more (all p<0.05 versus baseline). In addition, a 16% (nonsignificant) reduction in calories was observed in Intervention Group patients. Despite a significant reduction in fat intake, total fiber was almost completely unchanged, while the intake of RDA vitamins[56] was both adequate and unchanged compared to baseline for women in the Dietary Intervention Group.

VI. Correlates of Dietary Change

In addition to the self-reported dietary results outlined, objective parameters consistent with dietary changes were also observed. As shown in Table 2, weight change from entry was significantly different in the Intervention versus the Control Groups, with a difference in weight at 24 months of approximately 3.5 kilograms (p<0.05). The magnitude of the weight differences strongly suggests that major dietary alteration was initiated and maintained in the Intervention Group. Thus, the WINS dietary program to reduce fat intake successfully abrogated the weight increase commonly associated with therapeutic programs for women with resected breast cancer, especially those receiving systemic adjuvant chemotherapy. As previously seen, such gain in weight following systemic adjuvant treatment has been associated with an adverse clinical outcome. As such, prevention of weight increase may represent a separate endpoint with quality of life implications for women with resected breast cancer. If, as is likely, the weight increase is related to a change in eating pattern which adversely influences breast cancer growth and metastatic spread, prevention of the weight increase as demonstrated in this feasibility study may well benefit breast cancer patient outcome as well.

A further correlate of the reported dietary intake change was the change in estradiol levels noted in the subset of patients in the WINS trial who had baseline and follow-up blood samples analyzed for established levels. As reported by Rose and colleagues,[26,27] in a subset of 57 women on the WINS feasibility trial, estradiol decreased 20% from baseline to follow-up at 18 months (p<0.05) for breast cancer patients in the Dietary Intervention Group, whereas estradiol levels in the Dietary Control Group did not change. The Women's Health Trial (WHT) represents a dietary intervention study involving women without a breast cancer diagnosis but at increased risk for breast cancer development.[54] Published feasibility results for that study are closely comparable to that of the WINS Trial and are illustrated side by side in Table 4. As shown, the magnitude of the dietary fat intake reduction, the change in body weight, and the change in estradiol levels in the two populations are closely comparable.[24]

The only difference regarding secondary markers of dietary adherence between the WINS and WHT trials was in the area of serum cholesterol levels. In the Women's Health Trial, a 9% reduction in serum cholesterol was noted, consistent with that predicted by the Keys equation, in the Dietary Intervention Group.[54] In the Women's Intervention Nutrition Study, only a 4% reduction in serum cholesterol occurred in the group on reduced dietary fat.[2] However, the majority of women on the Women's Intervention Nutrition Study were also receiving tamoxifen, an agent known to lower serum cholesterol and alter lipid profile.[46,48] In fact, these qualities associated with tamoxifen use provide the basis for its evaluation as an agent to reduce cardiovascular risk.[47,48] However, information on the interaction between tamoxifen and reduction in dietary fat in changing lipid profile has not been reported. This area is receiving attention now by investigators from the Women's Intervention Nutrition Study.[57]

The data demonstrating self-reported reduction in dietary fat intake along with associated correlates supporting dietary change are substantial and are in close agreement with that reported for other trials. For example, Holm et. al.[7] reported in a European population of approximately 240 patients with resected breast cancer, comparable changes in dietary fat intake and patient weight occurred in a population with localized and resected breast cancer when randomized to enter a program aimed at reducing dietary fat. It appears that even in the complex environment following a diagnosis of breast cancer in women receiving conventional local and systemic therapy, reduction in dietary fat intake can be successfully implemented in a multi-site study.

VII. Women's Intervention Nutrition Study: Definitive Outcome Trial

Based on the feasibility information outlined above, a multi-site full-scale outcome study of the Women's Intervention Nutrition Study (WINS) has recently been initiated in the United States with support from the National Cancer Institute. The design is targeted to enter 2,000 patients with resected breast cancer (stage I-IIb) who are receiving conventional systemic adjuvant therapy and are randomized to reducing dietary fat or no such program. A similar trial is planned for implementation in Europe as well. Thus, definitive evidence regarding a potential influence of dietary fat intake on breast cancer growth and metastatic spread should be available in the near future.

Although the sample size of the full-scale WINS effort outline is based on a projected reduction in the relapse rate of breast cancer and on improvement in patient survival, information regarding the influence of lowering dietary fat intake on primary prevention of breast cancer may also be generated. There is a relatively high risk of cancer development in the opposite breast in patients with an established breast cancer diagnosis. In fact, a major impetus to the suggestion that tamoxifen may act as a preventive agent for breast cancer were observations that new cancer in the opposite breast was significantly lower in women on tamoxifen in adjuvant trials compared to those not receiving this agent.[46]

In summary, based on a sound rationale and feasibility trial support, full scale studies are now underway to address definitively the hypothesis relating dietary fat intake to clinical outcome of patients with early stage, resected breast cancer. The development of objective correlates of adherence to programs for reduction in dietary fat intake provides assurance that dietary change will be achieved in these efforts without sole reliance on self-reports of

Table 4. Effect of Dietary Fat Reduction Program on Dietary Intake, Estradiol, and Body Weight Change for Participants in WINS and the WHT

Parameters	WHT	WINS
Number	73	57
Follow-up	10-22 weeks	18 months
Dietary fat intake (daily fat grams)	−43% *	−47% *
Estradiol	−17% *	−20% *
Body Weight	−3.4 KG *	−3.8 KG *

* Baseline to follow-up (p<0.05)

WINS: Women's Intervention Nutrition Study.
WHT: Women's Health Trial.

Data summarized from References 24, 26 and 27.

dietary intake.[58] The outcome of these efforts will determine whether dietary counseling to reduce fat intake will become a routine component of breast cancer patient management in the future.

Acknowledgment

Partly supported by grant CA45502 from the National Cancer Institute, Bethesda, MD.

References

1. Chlebowski, R.T., D. Rose, I.M. Buzzard, G.L. Blackburn, W. Insull Jr., M. Grosvenor, R. Elashoff, and E.L. Wynder, Adjuvant dietary fat intake reduction in postmenopausal breast cancer management, *Breast Cancer Res. Treat.* 20:73 (1991).
2. Chlebowski, R.T., G.L. Blackburn, I.M. Buzzard, D.P. Rose, S. Martino, J.D. Khandekar, R.M. York, R.W. Jeffrey, R.M. Elashoff, and E.L. Wynder, For the Women's Intervention Nutrition Study. Adherence to a dietary fat intake reduction program in postmenopausal women receiving therapy for early breast cancer, *J. Clin. Oncol.* 11:2072 (1993).
3. Cohen, L.A., D.P. Rose, and E.L. Wynder, A rationale for dietary intervention in the treatment of postmenopausal breast cancer patients: an update, *Nutr. Cancer.* 19:1 (1993).
4. Chlebowski, R.T., D.W. Nixon, G.L. Blackburn, P. Jochimsen, E.F. Scanlon, W. Insull Jr., I.M. Buzzard, R. Elashoff, R. Butrum, and E. L. Wynder, A breast cancer Nutrition Adjuvant Study (NAS): protocol design and initial patient adherence, *Breast Cancer Res. Treat.* 10:21 (1987).
5. Boyar, A.P., D.P. Rose, J. R. Loughridge, A. Engle, A. Palgi, K. Laakso, D. Kinne, and E.L. Wynder, Response to a diet low in total fat in women with postmenopausal breast cancer: a pilot study, *Nutr. Cancer.* 11:93 (1988).
6. Boyd, N.F., M. Cousins, and V. Kriukov, A randomized controlled trial of dietary fat reduction: the retention of subjects and characteristics of drop outs, *J. Clin. Epidemiol.* 45:31 (1992).
7. Holm, L.-E., E. Nordevang, E. Ikkala, L. Hallstrom, and E. Callmer, Dietary intervention as adjuvant therapy in breast cancer patients: a feasibility study, *Breast Cancer Res. Treat.* 16:103 (1990).
8. Kristal, A.R., E. White, A.L. Shattuck, S. Curry, G.L. Anderson, A. Fowler, and N. Urban, Long-term maintenance of a low-fat diet: durability of fat-related dietary habits in the Women's Health Trial, *J. Am. Diet. Assoc.* 92:553 (1992).
9. Wynder, E.L., T. Kajitani, J. Kuno, J.C. Lucas Jr., A. de Palo, and J. Farrow, A comparison of survival rates between American and Japanese patients with breast cancer, *Surg. Gynecol. Obstet.* 117:196 (1963).
10. Sakamoto, G., H. Sugano, and W.H. Hartman, Stage-by-stage survival from breast cancer in the U.S. and Japan, *Jpn. J. Cancer.* 25:161 (1979).
11. Markita, M. and G. Sakamoto, Natural history of breast cancer among Japanese and Caucasian females, *Gan To Kagaku Ryoho* 17:1239 (1990).
12. Allen, D.S., R.D. Bulbrook, M.A. Chaudary, J.L. Hayward, M. Yoshida, S. Miura, and J.T. Murai, Recurrence and survival rates in British and Japanese women with breast cancer, *Breast Cancer Res. Treat.* 18:s131 (1991).
13. Morrison, A.S., C.R. Lowe, B. MacMahon, B. Ravnihar, and S. Yuasa, Some international differences in treatment and survival in breast cancer, *Int. J. Cancer* 18:269 (1976).
14. Lands, W.E.M., T. Hamazaki, K. Yamazaki, H.Okuyama, K. Sakai, Y. Goto, and V.S. Hubbard, Changing dietary patterns, *Am. J. Clin. Nutr.* 51:991 (1990).
15. Nomura A.M.Y., L. Le Marchand, L.N. Kolonel, and J.H. Hankin, The effect of dietary fat on breast cancer survival among Caucasian and Japanese women in Hawaii, *Breast Cancer Res. Treat.* 18:135s (1991).
16. Henderson, I.C., Rapporteur's report — Treatment and response, *Breast Cancer Res. Treat.* 18:157s (1991).

17. Early Breast Cancer Trialists' Collaborative Group, Systemic treatment of early breast cancer by hormonal cytotoxic or immune therapy, *Lancet* 339:1 (1992).

18. Willett, W.C., D.J. Hunter, M. Stampfer, G. Colditz, J.E. Manson, D. Spiegelman, B. Rosner, C.H. Hennekens, and F.E. Speizer, Dietary fat and fiber in relation to risk of breast cancer. An 8-year follow-up, *JAMA* 268:2037 (1992).

19. Howe, G.R., T. Hirohata, T.G. Hislop, J.M. Iscovich, J.-M. Yuan, K. Katsouyanni, F. Lubin, E. Marubini, B. Modan, T. Rohan, P.T. Toniolo, and Y. Shunzhang, Dietary factors and risk of breast cancer: combined analysis of 12 case-control studies, *J. Natl. Cancer Inst.*, 82:561 (1990).

20. Erickson, K.L. and N.E. Hubbard, Dietary fat and tumor metastasis, *Nutr. Rev.* 48:6 (1990).

21. Rose, D.P., J.M. Connolly, and C.L. Meschter, Effect of dietary fat on human breast cancer growth and lung metastasis in nude mice, *J. Natl. Cancer Inst.* 83:1491 (1991).

22. Rose, D.P. and J.M. Connolly, Influence of dietary fat intake on local recurrence and progression of metastases arising from MDA-MB-435 human breast cancer cells in nude mice after excision of the primary tumor, *Nutr. Cancer* 18:113 (1992).

23. Longcope, C., Relationships of estrogen to breast cancer, of diet to breast cancer, and of diet to estradiol metabolism, *J. Natl. Cancer Inst.* 82:896 (1990).

24. Prentice, R., D. Thompson, C. Clifford, S. Gorbach, B. Goldin, and D. Byar, Dietary fat reduction and plasma estradiol concentration in healthy postmenopausal women, *J. Natl. Cancer Inst.* 82:129 (1990).

25. Rose, D.P., A. Boyar, C. Cohen, and L.E. Strong, Effect of a low-fat diet on hormone levels in women with cystic breast disease. I. Serum steroids and gonadotropins, *J. Natl. Cancer Inst.* 78:623 (1987).

26. Rose, D.P., R.T. Chlebowski, J.M. Connolly, L.A. Jones, and E.L. Wynder, The effects of tamoxifen adjuvant therapy and a low-fat diet on serum binding proteins and estradiol bioavailability in postmenopausal breast cancer patients, *Cancer Res.* 52:5386 (1992).

27. Rose, D.P., J.M. Connolly, R.T. Chlebowski, I.M. Buzzard, and E.L. Wynder, The effects of a low-fat dietary intervention and tamoxifen adjuvant therapy on the serum estrogen and sex hormone-binding globulin concentrations of postmenopausal breast cancer patients, *Breast Cancer Res. Treat.* 27:253 (1993).

28. Shimizu, H., R.K. Ross, L. Bernstein, M.C. Pike, and B.E. Henderson, Serum oestrogen levels in postmenopausal women: comparison of American whites and Japanese in Japan, *Br. J. Cancer.* 62:451 (1990).

29. Wang, D.Y., T.J.A. Key, M.C. Pike, J. Boreham, and J. Chen, Serum hormone levels in British and rural Chinese females, *Breast Cancer Res. Treat.* 18:41 (1991).

30. Senie, R.T., P.P. Rosen, P. Rhodes, M.L. Lesser, and D.W. Kinne, Obesity at diagnosis of breast carcinoma influences duration of disease-free survival, *Ann. Intern. Med.* 116:26 (1992).

31. Bastarruchea, J., G.N. Hortobagyi, T.L. Smith, S.C. Kai, and A. U. Buydan, Obesity as an adverse prognostic factor for patients receiving adjuvant chemotherapy for breast cancer, *Ann. Intern. Med.* 119:18 (1993).

32. Camoriano, J.K., C.L. Loprinzi, J.N. Ingle, T.M. Therneau, J.E. Krook, and M.H. Veeder, Weight change in women treated with adjuvant therapy or observed following mastectomy for node-positive breast cancer, *J. Clin. Oncol.* 8:1327 (1990).

33. Chlebowski, R.T., J.M. Weiner, R. Reynolds, J. Luce, L. Bulcavage, and J.R. Bateman, Long term survival following relapse after 5-FU but not CMF adjuvant breast cancer therapy, *Breast Cancer Res. Treat.* 7:23 (1986).

34. Dixon, J., D.A. Moritz, and F.L. Baker, Breast cancer and weight gain: an unexpected finding, *Oncol. Nurs. Forum* 5:5 (1978).

35. Hoskin, P.J., S. Ashley, and J.R. Yarnold, Changes in body weight after treatment for breast cancer and the effect of tamoxifen, *Br. J. Cancer* 62 (Suppl XI):26 (1990).

36. Denmark-Wahnefried, W., E.P. Winer, and B.K. Rimer, Why women gain weight with adjuvant chemotherapy for breast cancer, *J. Clin. Oncol.* 11:1418 (1993).

37. Heasman, K.Z., J.H. Sutherland, J.A. Campbell, T. Elhakim, and N.F. Boyd, Weight gain during adjuvant chemotherapy for breast cancer, *Breast Cancer Res. Treat.* 5:195 (1986).

38. Gregorio, D., L. Emrich, S. Graham, J. Marshall, and T. Nemoto, Dietary fat consumption and survival among women with breast cancer, *J. Natl. Cancer Inst.* 75:37 (1985).

39. Holm, L.-E., E. Callmer, M.-L. Hjalmar, E. Lidbrink, B. Nilsson, and L. Skoog, Dietary habits and prognostic factors in breast cancer, *J. Natl. Cancer Inst.* 81:1218 (1989).

40. Holm, L.-E., E. Nordevang, M.-L. Hjalmar, E. Lidbrink, E. Callmer, and B. Nilsson, Treatment failure and dietary habits in women with breast cancer, *J. Natl. Cancer Inst.* 85:32 (1993).

41. Tretli, S., T. Haldorsen, and L. Ottestad, The effect of pre-morbid height and weight on the survival of breast cancer patients, *Br. J. Cancer* 62:299 (1990).

42. Henderson, M.M., Role of intervention trials in research on nutrition and cancer, *Cancer Res.* 52:2030s (1992).

43. Ewertz, M., S. Gillanders, L. Meyer, and K. Zedeler, Survival of breast cancer patients in relation to factors which affect the risk of developing breast cancer, *Int. J. Cancer* 49:526 (1991).

44. Rohan, T., J. Hiller, and A. McMichael, Dietary factors and survival from breast cancer, *Nutr. Cancer* 20:167 (1993).

45. Newman, S.C., A.B. Miller, and G.R Howe, A study of the effect of weight and dietary fat on breast cancer survival time, *Am. J. Epidemiol.* 123:767 (1986).

46. Kyogoku, S., T. Hirohata, Y. Nomura, T. Shigematsu, S. Takeshita, and T. Hirohata, Diet and prognosis of breast cancer, *Nutr. Cancer* 17:271 (1992).

47. Nayfield, S.G., J.E. Karp, L.G. Ford, F.A. Dori, and B.S. Kramer, Potential role of tamoxifen in prevention of breast cancer, *J. Natl. Cancer Inst.* 83:1450 (1991).

48. Love, R.R., P.A. Newcomb, D.A. Wiebe, T.S. Surawicz, V.C. Jordan, P.P. Carbone, and D.L. DeMets, Effects of tamoxifen therapy on lipid and lipoprotein levels in postmenopausal patients with node-negative breast cancer, *J. Natl. Cancer Inst.* 82:1327 (1990).

49. Buzzard, I.M., E.H. Asp, R.T. Chlebowski, A.P. Boyar, R.W. Jeffrey, D.W. Nixon, G.L. Blackburn, P.R. Jochimsen, E.F. Scanlon, W. Insull Jr., R.M. Elashoff, R. Butrum, and E.L. Wynder, Diet intervention methods to reduce fat intake: nutrient and food group composition of self-selected low-fat diets, *J. Am. Diet. Assoc.* 90:42 (1990).

50. Chlebowski, R.T., D.P. Rose, I.M. Buzzard, G.L. Blackburn, M. York, W. Insull Jr., J. Khandekar, S. Martino, R.M. Elashoff and E.L. Wynder, Dietary fat reduction in adjuvant breast cancer therapy: current rationale and feasibility issues, *Adjuvant Ther. Cancer* 6:357 (1990).

51. Nordevang, E., E. Ikkala, E. Callmer, L. Hallstrom, and L.-E. Holm, Dietary intervention in breast cancer patients: effects on dietary habits and nutrient intake, *Eur. J. Clin. Nutr.* 44:681 (1990).

52. Nordevang, E., E. Callmer, A. Marmur, and L.-E. Holm, Dietary intervention in breast cancer patients: effects on food choice, *Eur. J. Clin. Nutr.* 46:387 (1992).

53. Gorbach, S.L., A. Morrill-LaBrode, M.N. Woods, J.T. Dwyer, W.D. Selles, M. Henderson, W. Insull Jr., S. Goldman, D. Thompson, C. Clifford, and L. Sheppard, Changes in food patterns during a low-fat dietary intervention in women, *J. Am. Diet. Assoc.* 90:802 (1990).

54. Henderson, M.M., L.H. Kushi, D.J. Thompson, S.L. Gorbach, C.K. Clifford, W. Insull Jr., M. Moskowitz, and R.S. Thompson, Feasibility of a randomized trial of a low-fat diet for the prevention of breast cancer: dietary compliance in the Women's Health Trial Vanguard Study, *Prev. Med.* 19:115 (1990).

55. Hillner, B.E. and T.J. Smith, Efficacy and cost-effectiveness of adjuvant chemotherapy in women with node-negative breast cancer: a decision-analysis model, *N. Engl. J. Med.* 324:160 (1991).

56. Food and Nutrition Board: *Recommended Dietary Allowances (10th rev ed.).* Washington, D.C., National Academy of Sciences, 1989.

57. Chlebowski, R.T., G. Blackburn, J.P. Richie, and E. Wynder, Unanticipated effect of dietary fat intake reduction on HDL cholesterol in postmenopausal women receiving tamoxifen, *Proc. Am. Soc. Clin. Oncol.* 12:214 (1994).
58. Howat, P.M., R. Mohan, C. Champagne, C. Monlezun, P. Wozniak, and G.A. Bray, Validity and reliability of reported dietary intake data, *J. Am. Diet. Assoc.* 94:169 (1994).

Chapter 3

Dietary Fat Reduction as a Hypothesis for the Prevention of Postmenopausal Breast Cancer, and a Discussion of Hypothesis Testing Research Strategies

ROSS L. PRENTICE

I. Introduction

The hypothesis that a low fat diet may reduce the risk of human breast cancer has been promulgated for several decades. Experimental studies in rodent systems indicate specific roles for fat reduction and calorie restriction in inhibiting mammary tumor carcinogenesis.[1-3] International correlational studies suggest strong relationships between dietary fat and breast cancer incidence and mortality rates, particularly for postmenopausal women,[4-5] and there is supportive data from time trend and migrant studies.[5] Analytic epidemiologic studies, on the other hand, have sometimes[6] been interpreted as providing little or no support for an association of public health importance. Small scale human dietary intervention trials among postmenopausal women have documented changes in blood hormones,[7-8] but have not been of sufficient size to usefully examine dietary intervention effects on breast cancer incidence or survival. Hence the dietary fat and breast cancer hypothesis remains controversial even after decades of rather intensive epidemiologic study.

Here an updated summary of the epidemiologic data pertinent to the dietary fat and postmenopausal breast cancer hypothesis is provided, with an emphasis on the reliability of each specific type of study. Two facts emerge from this review: (i) available epidemiologic data are more consistent than may have been recognized by the authors of specific studies, and (ii) pertinent epidemiologic studies, individually and collectively, have substantial methodologic weaknesses, to the extent that the dietary fat and human breast cancer hypothesis has not yet been reliably tested. Several approaches to producing more reliable data and hypothesis tests are presented and discussed, including improved international studies that

---

Ross L. Prentice • Division of Public Health Sciences, Fred Hutchinson Cancer Research Center, 1124 Columbia Street, Seattle, Washington

Diet and Breast Cancer, Edited under the auspices of the American Institute for Cancer Research, Plenum Press, New York, 1994

include surveys of dietary factors and cancer risk factors in various populations worldwide, cohort studies in populations with widely varying dietary patterns using dietary assessment instruments having well-characterized measurement properties, and formal randomized controlled dietary intervention trials with breast cancer as an important outcome variable.

II. Brief Review of Epidemiologic Studies of Dietary Fat and Postmenopausal Breast Cancer

A. Aggregate Data Epidemiologic Studies

We have presented an analysis[5] of the incidence rates of selected cancers in relation to the per capita disappearance of fat, using cancer incidence data from Cancer Incidence in Five Continents, Volume V,[9] and nutrient supply data from the Food and Agriculture Organization of the United Nations[10] for 21 counties thought to have cancer incidence data representative of the country as a whole. Age-adjusted breast cancer incidence rates, for the years 1978-82, ranged from 57.9 per 100,000 in Japan to 257.4 per 100,000 in the United States.[5] Corresponding per capita fat kilocalories averages for the years 1975-77 ranged from 648 per day in Japan to 1,476 per day in the United States.[11] Simple linear regression analysis of log-incidence rates in per capita fat kilocalories for the 21 countries yields relative risk estimates (95% confidence intervals) for a 50% reduction from U.S. levels of total fat of 0.39 (0.27, 0.55). Hence if this simple analysis is accurate a 2½-fold reduction in postmenopausal breast cancer risk in the United States would eventually be expected to follow from a reduction by 50% in fat consumption in the United States. Such a reduction has been shown to be practical among middle aged and older women in extensive feasibility trials of low fat dietary intervention programs.[12-13]

Even though the regression model mentioned above appears to fit the breast cancer incidence rates, and it provides an explanation for 63% of the incidence rate variations among the 21 countries, there are a number of obvious concerns about this simple regression analysis. Firstly, the dietary data represent per capita estimates of nutrient supply, rather than of per capita nutrient consumption. Hence the per capita fat calorie estimates will be inflated by food wastage and by food consumption by household pets, for example. The per capita supply (disappearance) estimates are also not sex- or age-specific. Note, however, that the per capita fat disappearance estimates could be multiplied by a common positive constant (e.g. 0.5) toward bringing them into approximate agreement with the per capita consumption estimates without affecting the relative risk (0.39) mentioned above, its confidence interval, or the percentage of incidence rate variation explained. It can also be commented that these nutrient disappearance estimates are one of the few sources of dietary information that do not depend on individual dietary recall, and that the correlation across countries of nutrient disappearance estimates and nutrient consumption estimates from the dietary survey of individuals is rather high.[14]

Perhaps the more important concern with the regression analysis mentioned above relates to the absence of control for factors that may confound these aggregate analyses. Adding non-fat calories to the regression analysis[5] changes the relative risk estimate for a 50% reduction in fat only from 0.39 to 0.36, while adding per capita gross national product changes the relative risk estimate from only 0.39 to 0.40. Neither of these factors relate significantly to breast cancer incidence in the presence of fat calorie disappearance. In closely related analyses[11] none of protein, alcohol and carbohydrate calories supply; retinol and beta-carotene supply; average height; average weight; or average age at menarche materially affected the fat calorie relative risk estimate, and none of these factors related significantly to breast cancer incidence in the presence of per capita fat calorie supply. Nonetheless such analyses represent a somewhat limited attempt at confounding factor control, and there is good reason to conduct a well controlled and standardized aggregate data study of diet and cancer, as will be discussed below.

One way to examine the reliability of the relative risk estimate mentioned above is to examine its consistency with relative risk estimates from other data sources. For example a regression[5] of the change in log-breast cancer incidence rates among women of ages 55-69 between Volume II (about 1965) and Volume V (about 1980) on the corresponding change in per capita fat calorie disappearance between 1961-63 and 1975-77 for a subset of 10 of the 21 countries, gives a relative risk estimate for a 50% reduction from U.S. levels of fat calories of 0.45, with a corresponding 95% confidence interval of (0.19, 1.10). This estimate allows for a general increase in breast cancer incidence rates, estimated as 1.25, in addition to that attributable to the change in fat calories. Hence, this relative risk estimate based on within country information agrees well with that previously given, which is based on between country comparisons.

Consistency of the results of international regression analyses with the results of cohort and case-control studies is more difficult to assess, owing to the profound effect that measurement error in individual dietary assessment may have on the results of such analytic epidemiologic studies.

B. Analytic Epidemiologic Studies

Various instruments, including food records, food frequency questionnaires, dietary recalls and dietary histories have been used in an attempt to assess the nutrient content of the diets of individual study subjects. Reliability studies and so-called validation studies make it clear that such measurements include considerable error with correlations based on repeat application of an instrument at different times, or based on the application of two or more instruments at a specific time, tending to be in the range 0.2 to 0.6, depending on the nutrient and the instrument (e.g., 15). The fundamental limitation to reliable relative risk estimation from diet and disease cohort and case-control studies arises from the fact that it is not possible to conduct a real validation study in which true nutrient intakes are available for comparison to estimated intakes, among persons under free living conditions. Hence it is necessary to make strong modelling assumptions in any attempt to relate disease risk to true, but unobservable, nutrient consumption. A classical measurement model would assume that a true nutrient intake z, say average daily grams of fat over a specified time period, gives rise to a corresponding estimate x via

$$x = \alpha z + \varepsilon \tag{1}$$

where z and the random measurement error ε are assumed to be independently normally distributed and $\alpha > 0$ is a scaling factor. If, as in the previous subsection, breast cancer relative risk is an exponential function, $\exp(z\beta)$, of daily fat intake z then the relative risk as a function of the measured fat intake x can be written $\exp(x\rho\beta)$, where ρ is the ratio of the covariance of x and z to the variance of x. For example, if one assumes the grams of fat estimate from the 28 days of food records in the 'validation' study of Willett and colleagues[15] are without error and defines x as the corresponding food frequency grams of fat estimate, one can estimate $\rho=0.274$, the ratio of the food-record and food-frequency sample covariance to the food-frequency sample variance. The international breast cancer analysis mentioned above gives an estimated regression coefficient of $\beta=0.0275$ upon rescaling the food disappearance fat grams values for each country so that the U.S. value equals the sample mean of the food-record daily grams of fat in Willett et al.[15] Since the food frequency instrument was used for dietary assessment in the nurses cohort study,[6,16] one can use this result to project breast cancer relative risks across quintile medians in the nurses study. Upon standardizing so that the relative risk is unity in the lowest fat intake quintile, one obtains international data projected relative risks[5] of 1, 1.07, 1.15, 1.26 and 1.53 across food frequency fat intake quintiles in the nurses study. These projections contrast with the estimated relative risks of 1, 0.75, 0.64, 0.84 and 0.89 (p=0.22) among postmenopausal women in the original fat and breast cancer report from the nurses cohort.[16] The later report[6] (774 postmenopausal breast cancer cases) presented results only by calorie adjusted food-frequency fat intake quintiles, reporting relative risk estimates of 1, 0.89, 1.00, 0.95 and 0.91 among postmenopausal women. In fact, the discrepancy between expectation based on strong international correlations and the absence of any hint of positive relationship in this, the largest of the cohort studies is a major, if not the major, reason that the dietary fat and breast cancer hypothesis remains controversial.

Other analytic epidemiologic studies appear to be more consistent with the strong international regression analysis. Specifically, Howe and colleagues[17] carried out a joint analysis of the raw data from 12 case-control studies, that included 4,247 cases and 6,095 controls, about two-thirds of whom were postmenopausal. This analysis indicated a highly significant positive association between breast cancer risk and estimated daily grams of fat (p<0.0001) among postmenopausal women. The estimated breast cancer relative risk[5] by quintiles, defined by a Canadian case-control study, were 1, 1.20, 1.24, 1.24 and 1.46. These estimates appear to be quite consistent with the international regression analysis projections given above, though undoubtedly such projections should depend somewhat on the properties of the dietary assessment instruments used in the 12 case-control studies.

There are three other sizable cohort studies of fat consumption and breast cancer. Kushi et al.[18] give relative risk estimates of 1, 1.17, 1.25 and 1.38 (p=0.18) across estimated fat intake quartiles (459 postmenopausal cases), Howe et al.[19] give relative risk estimates of 1, 0.73, 0.98 and 1.30 (p=0.05) across fat intake quartiles (519 pre- and postmenopausal cases). Relative risk estimates of 1, 1.00, 1.34, 1.22 and 1.08 (p=0.32) across fat intake quintiles (471 postmenopausal cases), were reported by van den Brandt et al.[20] where each set of relative risk estimates has been adjusted for calories and possibly other factors. These studies may be consistent with the international regression analysis, but they may also be consistent with no dietary fat and breast cancer association whatsoever.

C. Other Studies

Tominaga[21] presented data indicating that age-adjusted breast cancer incidence rates among Japanese migrants (first and later generations) to the United States were 3.5 times those among Japanese in Japan. In comparison, upon making a reasonable assumption[5] concerning the change in dietary fat following migration, our international regression analysis predicts a breast cancer (ages 55-69) relative risk of 2.9 among Japanese women in the United States versus Japanese in Japan. A recent breast cancer case-control study among younger women indicates that Japanese migrants to the United States experience an appreciable increase in breast cancer risk within a decade of migration.[22]

The per capita daily disappearance of polyunsaturated, saturated, and monounsaturated fat in the United States was 17, 59 and 54 grams, respectively, in 1945, as compared to 31, 58 and 63 grams in 1980.[23] International regression analysis[5] of breast cancer incidence among women ages 55-69 on estimates of polyunsaturated, saturated and monounsaturated fat disappearance indicates positive associations for both polyunsaturated and saturated fat, with polyunsaturated fat having the stronger association. There was no association with monounsaturated fat supply. Regression analyses can be used to project a breast cancer increase of 47% corresponding to the fat disappearance changes between 1945 and 1980.[24] In comparison breast cancer incidence in the United States increased by an estimated 42% between 1947-50 and 1983-84.

It is of interest, in view of well-established relationships between hormone-dependent events in a woman's life and breast cancer risk, to note that plasma estradiol concentrations among postmenopausal women not taking exogenous estrogens dropped by an average of 17% (p<0.001) within 10-22 weeks of undertaking a low fat diet as part of the Women's Health Trial.[8] A reduction by this amount provides an explanation for a major component of the international variation in breast cancer incidence rates. Other investigators[7,25] have also documented reductions in blood estrogens among post- and premenopausal women undertaking a low fat eating pattern.

III. Interpretation of Studies Relating Dietary Fat to Postmenopausal Breast Cancer Incidence

Two very different interpretations of the data relating fat consumption to postmenopausal breast cancer risk have been given. One interpretation[5] notes that even a crude measure of per capita fat supply provides an explanation for 63% of the 5-fold international variation in postmenopausal breast cancer rates. These international correlational analyses suggest that a 50% reduction in fat consumption could eventually reduce postmenopausal breast cancer rates in the United States by 50-75%, giving one of the few practical possibilities for substantially reducing the public health impact of this dreaded disease. Lowering dietary fat can also be projected to induce major reductions in colorectal, ovarian, endometrial, and prostate cancer rates, and a reduction in coronary heart disease, making a reliable study of the benefits and risks of a low fat eating pattern among the highest of research priorities. Also international correlation analyses appear to be consistent with the results of various other types of studies, including national and international time trend studies of disease rates, migrant studies, case-control studies, and certain cohort studies, with the notable exception of the nurses cohort study. Importantly, the international correlational analyses are supported qualitatively by the results of animal feeding trials, including the finding of a possible substantial role for polyunsaturated fat consumption[2] (see also the contribution by Freedman and Clifford in this volume).

The second interpretation argues[6] that "the totality of evidence does not support the judgment of a causal association" between fat consumption and postmenopausal breast cancer risk. Specifically it is argued[6,26,27] that international comparisons and time trends in disease rates may be incorrect because of possible confounding and because they are based on food disappearance rather than food consumption data; that certain human intervention trials that have noted reductions in blood estrogen concentrations among women undertaking a low fat diet may be misleading because they have not studied corresponding changes among control women; that the positive association observed in the pooled analysis of 12 case-control studies may be due to "bias inherent in the retrospective assessment of diet";[6] that collectively the cohort studies to date "have found little or no evidence of any positive relation between fat intake and breast cancer risk,"[16] and finally that the "inability to demonstrate the positive relation suggested by international correlations in the nurses study cannot be explained by imprecision in the measurement of fat intake."[6]

Counter arguments can be presented to each of these viewpoints. For example, one dietary intervention trial[8] showing a significant average 17% reduction in blood estradiol among dietary intervention women involved follow-up times of only 10-22 weeks, making it implausible that such reductions would have occurred in the absence of dietary intervention; that a retrospective ascertainment of dietary factors among women in the Canadian cohort study cited above[19] failed to provide evidence for recall bias in estimating fat consumption;[28] and that aggregate data studies appear to be much less susceptible to the effects of random measurement error in exposure assessment than are cohort studies.[29]

Nevertheless, both viewpoints mentioned are being held, pointing to the need for more reliable studies to assess the dietary fat and breast cancer hypothesis.

Before discussing the types of studies necessary to advance knowledge concerning diet and cancer, it is worth commenting on the claim[6] that the nurses study results cannot be explained by "imprecision in the measurement of fat intake." In fact, better understanding of the possible impact of measurement error in dietary assessment seems to be key to the rational weighting of the evidence from various sources and to the choice of study designs and approaches for future research.

The claim just mentioned by Willett and colleagues is based on the simple measurement error model (1). But this very simplistic model assumes that any bias in estimates of fat consumption and the measurement error variance in fat consumption estimates are independent of study subject characteristics. In general, the absence of true validation studies makes it impossible to assess such assumptions. However, there is increasing evidence from studies using the doubly labelled water method to measure calorie expenditure which indicates that obese persons under-report calorie consumption by 25-50%.[30-32] The degree of under-reporting of fat calories may be even greater than for non-fat calories. It should be noted that a departure from the simple measurement model (1) that allows the measurement properties of fat intake estimates to differ between obese and non-obese women, can dramatically change the shape of the relative risk function across categories of measured fat intake. Suppose that the postmenopausal breast cancer relative risk can be written $\exp(z\beta)$, as before, where z is an individual's actual fat consumption, and β is specified so that a 50% reduction from the mean U.S. fat consumption projects a relative risk of 0.39, as was estimated in our international regression analysis.[5] Suppose now that 20% of a postmenopausal cohort of women are defined as obese, according to some suitable criterion; that the mean fat consumption for obese women is 25% higher than that in non-obese women; that the standard deviation of fat intake is equal in obese and non-obese women and equals 12.5% of the mean intake for non-obese women; that the measurement error standard deviation equals that of the true fat intake standard deviation for both obese and non-obese women; and finally that obesity is not an independent breast cancer risk factor, given true fat consumption. A simple exercise then shows the breast cancer relative risk from smallest to largest quintile median of estimated fat intake to be 1, 1.16, 1.30, 1.47 and 1.80, provided the measured fat intake estimates are unbiased; that is, assuming that $\alpha=1$ in model (1) for both obese and non-obese women. Hence, under these measurement assumptions there would be a clear ability to detect a trend in breast cancer rates across the estimated fat intake categories in the cohort study. Now retain the above assumptions but suppose that fat calories are under-reported by 50% among the obese women in the cohort (i.e., $\alpha=0.5$ in model (1) for obese women). The projected breast cancer relative risks across quintile medians of estimated fat intake now become 1, 0.80, 0.79, 0.84 and 0.95. Hence, a plausible degree of under-reporting in a small subset of the cohort has completely eliminated a strong positive association of risk with true fat intake and replaced it by a slightly negative association between risk and estimated fat intake. This exercise makes it clear that the measurement properties of a dietary assessment instrument must be well known in order to interpret corresponding analytic epidemiologic study results. It indicates that dietary assessment measurement issues may be at the heart of the differences of opinion surrounding the dietary fat and postmenopausal breast cancer hypothesis. Additional exercises of this type show that the shape of the projected relative risk function can depend dramatically on differences in the variance of measurement errors, as between obese and non-obese women, even if fat intake estimates are unbiased in both groups. They can depend markedly on plausible departures from the assumption that food frequency and food record measurement errors are uncorrelated. Also projected relative risks may be quite different across diseases (e.g., breast cancer and colon cancer) if such diseases depend differentially on the components of fat intake (e.g., saturated and polyunsaturated) which themselves have differing measurement properties.

IV. Future Research

Despite decades of intensive epidemiologic study, the dietary fat and postmenopausal breast cancer hypothesis has not been reliably tested. International comparisons of disease rates involve crude dietary data and may be subject to serious confounding. The results of analytic epidemiologic studies may be dominated by measurement errors in individual dietary assessment. It is then logical to ask how the human research agenda can proceed to test the hypothesis that a low fat eating pattern would have a major public health benefit in the United States and in other developed countries.

One important element of a future research agenda could be the conduct of a high quality international study of diet and cancer. Our group in Seattle is in the process of piloting a study which will eventually involve the surveying for dietary factors and other cancer risk factors of a few hundred randomly selected persons in the coverage areas of 30 or more good quality cancer registries worldwide. Such a study could provide good confounding factor control and estimated food consumption rather than food supply data. Our methodologic work[29,33] indicates that much of the pertinent diet and cancer information arises from between rather than within populations, and that suitable relative risk estimation procedures can be developed to extract such information. Furthermore these exercises indicate that such aggregate data analyses will be almost completely unaffected by classical measurement errors (1) in dietary assessment.

There is also a need for additional good quality analytic epidemiologic studies of diet and postmenopausal breast cancer. Studies that aim to include women with a broad range of dietary habits seem particularly well motivated. However, a principal focus in current and future cohort and case-control studies needs to be on the quality of dietary assessment. Much work is evidently required to develop methods for assessing the properties of the dietary assessment instruments to be used in a cohort or case-control study, thereby allowing more interpretable analytic study results.

Finally, the public health potential of a low fat eating pattern is such that a full-scale dietary intervention trial in postmenopausal women is strongly justified. Such a trial, to involve the randomization of 48,000 postmenopausal women in the age range 50-79 is currently underway as a part of the NIH's Women's Health Initiative. Our group serves as Clinical Coordinating Center and a Vanguard Clinical Center for this Initiative. This trial of dietary modification will focus on the prevention of breast cancer, colorectal cancer and coronary heart disease. It involves a dietary intervention program developed in previous and ongoing extensive feasibility studies.[12,34] Note that accurate individual dietary assessment will not be necessary to test reliably whether or not this low fat eating program can reduce the risk of breast cancer and other diseases in the Women's Health Initiative. However, such assessments would enter any effort to disentangle the effects of fat reduction from those of other nutritional changes incurred by the dietary modification program. Such efforts, though important, seem quite secondary to the identification of dietary changes that can improve the health of U.S. women.

In summary, there is considerable human data pertinent to the hypothesis that dietary fat reduction can substantially reduce the incidence of breast cancer among postmenopausal women. However, the design of each reported study is subject to fundamental criticisms, to the extent that this hypothesis has by no means been tested. It is essential to proceed with additional studies using stronger designs and procedures to advance knowledge in this critically important area of public health.

Acknowledgment

This work was supported by grant CA53996 from the National Institutes of Health.

References

1. Tannenbaum, A., Genesis and growth of tumors. III. Effects of a high fat diet, *Cancer Res.* 2:468 (1942).
2. Newberne, P.M., T.E. Shrager, and M.W. Conner, Experimental evidence on the nutritional prevention of cancer, *in: Nutrition and Cancer Prevention: Investigating the Role of Micronutrients*, T.E. Moon and M.S. Micozzi, eds., Marcel Dekker, New York (1989).
3. Freedman, L., C. Clifford, and M.W. Messina, Analysis of dietary fat, calories, body weight, and the development of mammary tumors in rats and mice: a review, *Cancer Res.* 50:5710 (1990).
4. Carroll, K.K., E.B. Gammal, and E.R. Plunkett, Dietary fat and mammary cancer, *Can. Med. Assoc. J.* 98:590 (1968).
5. Prentice, R.L. and L. Sheppard, Dietary fat and cancer: consistency of the epidemiologic data, and disease prevention that may follow from a practical reduction in fat consumption, *Cancer Causes Control* 1:81 (1990).
6. Willett, W.C., D.J. Hunter, M.J. Stampfer, G. Colditz, J.E. Manson, D. Spiegelman, B. Rosner, C.H. Hennekens, and F.E. Speizer, Dietary fat and fiber in relation to risk of breast cancer. An 8-year follow-up, *JAMA* 268:2037 (1992).
7. Boyar, A.P., D.P. Rose, J.R. Loughridge, A. Engle, A. Palgi, K. Laakso, D. Kinne, and E.L. Wynder, Response to a diet low in total fat in women with postmenopausal breast cancer. A pilot study, *Nutr. Cancer* 11:93 (1988).
8. Prentice, R., D. Thompson, C. Clifford, S. Gorbach, B. Goldin, and D. Byar, Dietary fat reduction and plasma estradiol concentration in healthy postmenopausal women, *J. Natl. Cancer Inst.* 82:129 (1990).
9. Muir, C., J. Waterhouse, T. Mack, J. Powell, and S. Whelan (eds.) *Cancer Incidence in Five Continents*, Vol. V, WHO, IARC, Lyon (1987).
10. Food and Agriculture Organization of the United Nations, *Food Balance Sheets 1975-1977 Average*, FAO, Rome (1980).
11. Prentice, R.L., F. Kakar, S. Hursting, L. Sheppard, R. Klein, and L.H. Kushi, Aspects of the rationale for the Women's Health Trial, *J. Natl. Cancer Inst.* 80:802 (1988).
12. Insull, W. Jr., M.M. Henderson, R.L. Prentice, D.J. Thompson, C. Clifford, S. Goldman, S. Gorbach, M. Moskowitz, R. Thompson, and M. Woods, Results of a feasibility study of a low-fat diet, *Arch. Intern. Med.* 150:421 (1990).
13. Boyd, N.F., M. Cousins, M. Beaton, V. Kriukov, G. Lockwood, and D. Tritchler, Quantitative changes in dietary fat intake and serum cholesterol in women: results from a randomized, controlled trial, *Am. J. Clin. Nutr.* 52:470 (1990).
14. Sasaki, S. and H. Kesteloot, Value of Food and Agriculture Organization data on food-balance sheets as a data source for dietary fat intake in epidemiologic studies, *Am. J. Clin. Nutr.* 56:716 (1992).
15. Willett, W.C., L. Sampson, M.J. Stampfer, B. Rosner, C. Bain, J. Witschi, C.H. Hennekens, and F. Speizer, Reproducibility and validity of a semiquantitative food frequency questionnaire, *Am. J. Epidemiol.* 122:51 (1985).
16. Willett, W.C., M.J. Stampfer, G.A. Colditz, B.A. Rosner, C.H. Hennekens, and F.E. Speizer, Dietary fat and the risk of breast cancer, *N. Engl. J. Med.* 316:22 (1987).
17. Howe, G.R., T. Hirohata, T.G. Hislop, J.M. Iscovich, J.-M. Yuan, K. Katsouyanni, F. Lubin, E. Marubini, B. Modan, T. Rohan, P. Toniolo, and Y. Shunzhang, Dietary factors and risk of breast cancer: combined analysis of 12 case-control studies, *J. Natl. Cancer Inst.* 82:561 (1990).
18. Kushi, L.H., T.A. Sellers, J.D. Potter, C.L. Nelson, R.G. Munger, S.A. Kaye, and A.R. Folson, Dietary fat and postmenopausal breast cancer, *J. Natl. Cancer Inst.* 84:1092 (1992).

19. Howe, G.R., C.M. Friedenreich, M. Jain, and A.B. Miller, A cohort study of fat intake and risk of breast cancer, *J. Natl. Cancer Inst.* 83:336 (1991).

20. van den Brandt, P.A., P. van't Veer, R.A. Goldbohm, E. Dorant, A. Volovics, R.J.J. Hermus, and F. Sturmans, A prospective cohort study on dietary fat and the risk of postmenopausal breast cancer, *Cancer Res.* 53:75 (1993).

21. Tominaga, S., Cancer incidence in Japanese in Japan, Hawaii, and Western United States, *Monogr. Natl. Cancer Inst.* 69:83 (1985).

22. Ziegler, R.G., R.N. Hoover, M.C. Pike, A. Hildeshein, A.M.Y. Nomura, D.W. West, A.H. Wu-Williams, L.N. Kolonel, P.L. Horn-Ross, J.F. Rosenthal, and M.B. Hyer, Migration patterns and breast cancer risk in Asian-American women, *J. Natl. Cancer Inst.* 85:1819 (1993).

23. United States Department of Agriculture, Nutrient content of the U.S. food supply, *in: Human Nutrition Information Service Admin. Report 299-21*, Hyattsville (1988).

24. Prentice, R.L., Rationale, feasibility, and design of a low fat diet intervention trial among postmenopausal women, *in: Macronutrients: Investigating their Role in Cancer*, M.C. Micozzi and T.E. Moon, eds., Marcel Dekker, New York (1992).

25. Rose, D.P., A.P. Boyar, C. Cohen, and L.E. Strong, Effect of a low fat diet on hormone levels in women with cystic breast disease. I. Serum steroids and gonadotropins, *J. Natl. Cancer Inst.* 78:623 (1987).

26. Willett, W., *Nutritional Epidemiology*, Oxford University Press, New York (1990).

27. Willett, W. and M.J. Stampfer, Dietary fat and cancer: another view, *Cancer Causes Control* 1:103 (1990).

28. Friedenreich, C.M., G.R. Howe, and A.B. Miller, The effect of recall bias on the association of calorie-providing nutrients and breast cancer, *Epidemiology* 2:424 (1991).

29. Prentice, R.L. and L. Sheppard, Aggregate data studies of disease risk factors, *Biometrika* (1994). (In Press).

30. Lichtman, S.W., K. Pisarska, E.R. Berman, M. Pestone, H. Dowling, E. Offenbacker, H. Weisel, S. Heshka, D.E. Matthews, and S.B. Heymsfield, Discrepancy between self-reported and actual calorie intake and exercise in obese subjects, *N. Engl. J. Med.* 327:1893 (1992).

31. Bandini, L.G., D.A. Schoeller, H.N. Cyr, and W.H. Dietz, Validity of reported energy intake in obese and nonobese adolescents, *Am. J. Clin. Nutr.* 52:421 (1990).

32. Livingstone, M.B.E., A.M. Prentice, J.J. Strain, W.A. Coward, A.E. Black, M.E. Barker, P.G. McKenna, and R.G. Whitehead, Accuracy of weighted dietary records in studies of diet and health, *Br. Med. J.* 300:708 (1990).

33. Sheppard, L. and R.L. Prentice, On the reliability and precision of within and between population estimates of relative risk parameters, *Biometrics* (1994). (In Press).

34. Henderson, M.M., L.H. Kushi, D.J. Thompson, S.L. Gorbach, C.K. Clifford, W. Insull Jr., M. Moskowitz, and R.S. Thompson, Feasibility of a randomized trial of a low-fat diet for the prevention of breast cancer: dietary compliance in the Women's Health Trial Vanguard Study, *Prev. Med.* 19:115 (1990).

Chapter 4
Hormone Studies and the Diet and Breast Cancer Connection

BARRY R. GOLDIN and SHERWOOD L. GORBACH

I. Introduction

The incidence rates for breast cancer vary widely in different parts of the world. The disease is more common in North America, Australia and Western Europe, relative to South and Central America, Asia and Africa.[1-3] Epidemiologic studies which relate per capita consumption of various dietary constituents to international variations in breast cancer incidence provide the most consistent evidence for an association between diet and breast cancer.[4-7] A significant positive association between total fat consumption and breast cancer incidence or mortality was found in all seven studies in which total dietary fat was reported. Among other dietary components that were compared to breast cancer rates, positive correlations with total calories, meat, sugar and specific fatty foods and negative correlations with cereals, beans, rice, maize and pulses were observed.[7]

Time trend and migrant studies have also revealed that changes toward a "Westernized" diet result in an increase in the incidence of breast cancer.[8] An example comes from the data on Japanese migrants to California which showed that the premenopausal Japanese migrant women have almost the same incidence of breast cancer as that observed in age-matched Caucasian women.[9]

II. Cohort Studies

Cohort studies show conflicting results with regard to the association of fat and other dietary constituents with breast cancer. In a large study in which 89,538 nurses were followed for more than eight years, no association was found between fat or fiber consumption and breast cancer incidence.[10,11] Phillips and Snowden[12] did a study of Seventh Day Adventists in California and found no relationship between the frequency of meat consumption and the mortality rate from breast cancer. In contrast, cohort studies in Japan have shown a relative risk for breast cancer of 3.83 for women eating meat almost daily versus

---

Barry R. Goldin and Sherwood L. Gorbach ● Department of Community Health, Tufts University School of Medicine, Boston, Massachusetts

Diet and Breast Cancer, Edited under the auspices of the American Institute for Cancer Research, Plenum Press, New York, 1994

women consuming meat less than once a week. There was a 2.86 and 2.10 relative risk for women eating eggs or butter almost daily, respectively, versus women eating these products less than once a week.[13]

III. Case-Control Studies

The data from case-control studies generally have shown an association of fat and other food products with the incidence or mortality from breast cancer. In a study involving 77 breast cancer cases and controls in the Seventh Day Adventist population, Phillips[14] found a positive association with relative risks ranging from 1.6 to 2.6 for fried food, fried potatoes, hard fat, dairy products (except milk), and white bread. A study in Canada on 400 breast cancer cases and 400 controls, originally conducted by Miller et al.[15] was re-analyzed by Howe et al.[16] who observed a significant risk for premenopausal women with increasing consumption of saturated fat (p=0.02). Lubin et al.[17] did another Canadian study with 577 cases and 826 controls and found that higher risk for breast cancer was associated with elevated consumption of beef, pork, butter and sweet desserts. Among premenopausal Singapore Chinese women a reduced risk of breast cancer was associated with high intakes of soya protein, β-carotene and polyunsaturated fats, but there was an increased risk with high intakes of total protein and red meat.[18]

Howe et al.[19] performed a meta-analysis of twelve case-control studies on diet and breast cancer. They found a significant direct association between breast cancer risk and saturated fat intake in postmenopausal women. An inverse risk of developing breast cancer was associated with several dietary indicators of high fruit and vegetable intake; vitamin C intake showed the most consistent protective effect. From these data it was concluded that dietary modification in North American women would result in a decline of 24% and 16% in breast cancer incidence for postmenopausal and premenopausal women, respectively.

In total, the epidemiological studies indicate that diet has a significant influence on breast cancer incidence, although this effect may involve various food constituents acting by different mechanisms.

IV. Diet and Sex Hormones

Diet-induced alterations in sex hormone metabolism is one mechanism by which diet may influence the process of mammary tumorigenesis. A number of observations have implicated hormones in the etiology of breast cancer.[20,21] Age above 30 years at first-term pregnancy, or nulliparity, increases relative risk (RR) (RR 2.0-4.0); early menarche and/or late menopause increases risk (RR 1.1-1.9); bilateral oophorectomy prior to 35 years of age lowers risk; use of oral contraceptives in young women (<30 years) appears to increase risk slightly; women over 35 years of age given estradiol for more than 6 years have an increased risk (RR 1.8) and women taking estrogen-progestin replacement therapy for more than 6 years have an increased risk (RR 4.4); postmenopausal women with abdominal adiposity have a two-fold greater risk than controls (adipose tissue is the major source of estrogens in postmenopausal women); and drugs and surgery that interfere with endogenous estrogen are useful in the treatment of breast cancer.

There is evidence that diet alters estrogen concentrations, pharmacokinetics and metabolism in healthy women.[22-27] Armstrong et al.[22] found differences in urinary estrogen levels as well as in sex hormone-binding globulin (SHBG) and prolactin levels between American vegetarians and omnivores. In a small study of 14 premenopausal Seventh-Day Adventists (SDA) and 9 premenopausal omnivores, plasma levels of estrone and estradiol were lower in the vegetarians, while a positive correlation was found between the concentration of plasma estradiol and dietary consumption of linoleic acid.[26] Rose et al.[27] reduced the fat intake from 35% to 21% in 16 premenopausal women with cystic breast disease. After three months on the low-fat diet, significant reductions were noted in serum estrone, estradiol and total estrogen. In our laboratory, healthy premenopausal women were switched from an American diet (40% fat, 12 gms of fiber per day) to a low fat (25%), high fiber diet (40 gms). A significant reduction in plasma estrone sulfate was observed after 2 months on the low fat, high fiber diet.[25] The pilot study of the Women's Health Trial (WHT) involved 73 healthy postmenopausal women who ate a low-fat diet for 10 to 22 weeks; significant reductions (17%) were noted in serum estradiol levels.[28]

V. Dietary Fiber

Dietary fiber also influences estrogen metabolism by several mechanisms. Fiber can decrease the enterohepatic circulation of estrogens by binding to estrogen in the intestine, thereby increasing fecal excretion at the expense of reabsorption from the intestine. Fiber increases fecal weight, which has been directly correlated with total fecal excretion of estrogen. Dietary fiber is derived from plant foodstuffs which include phytocompounds; some phytocompounds have weak estrogenic and appreciable anti-estrogenic activities. Farnsworth *et al.*[28] have reviewed the potential value of plants as a source of antifertility agents. The isoflavanoids, which are also called phytoestrogens, are the best known example of hormone-related plant compounds. Isoflavonic phytoestrogens were responsible for infertility in Australian sheep which grazed on formonentin-containing clover.[29,30] Formonentin is converted by intestinal bacteria to daidzein, equol and o-desmethylangolensin, which have all been identified in human urine.[31] Equol is considered responsible for the infertility syndrome in animals,[30] but humans eating soy products have elevated urinary levels of this phytoestrogen.

A second class of plant anti-estrogenic compounds, the plant "lignans", is excreted in urine of humans at levels greater than those of steroidal estrogens.[32] The two major lignans thus excreted are enterolactone and enterodiol, formed by intestinal bacteria from the plant lignan precursors, secoisolaricirestinal and matairesinol, which also occur at low levels in human urine.[31] Lariciresinol and isolaricinesinol, two other plant lignans in human urine, do not lead to enterolactone or enterodiol.[31] Urinary lignan levels correlate with consumption of fruits, berries, legumes, grain fiber and vegetable fiber.[33] These compounds appear to bind weakly, if at all, to type I estrogen receptors.[34] To summarize, the isoflavanoids have weak estrogenic activity and strong anti-estrogenic activity, while the lignans have weak estrogenic activity and moderate anti-estrogenic properties.

VI. Phytoestrogens and Breast Cancer

Phytoestrogens are implicated as anti-tumor agents, for these biphenolic plant compounds inhibit the growth of tumor cells, reduce steroid aromatase, angiogenesis, and DNA topo-isomerase II tyrosine kinase, and elevate sex hormone binding globulin in plasma.[32-37]

Adlercreutz and co-workers[36-39] compared the urinary lignan and phytoestrogen concentrations of populations eating different diets. Strict vegetarians (macrobiotics) had elevated levels of urinary lignans and phytoestrogens when compared to omnivores. Lacto-ovovegetarians also had elevated levels of lignans and phytoestrogens when compared to omnivorous women, but their concentrations were significantly lower than those of macrobiotics. Japanese women had elevated urinary phytoestrogens when compared to omnivores, but not higher lignans. Postmenopausal breast cancer patients had the lowest levels of urinary lignans and phytoestrogens when compared to any other group. Chimpanzees had high levels of urinary lignans and phytoestrogens, as would be predicted from their high consumption of plant products.[39] Based on these studies Adlercreutz and co-workers[39] concluded that the concentrations of anti-estrogenic plant compounds in the urine are correlated with plant food consumption. Populations at low risk for cancer (e.g. Japanese and vegetarians) have correspondingly higher urinary excretion of plant anti-estrogens. In this regard Lee *et al.*[18] reported that consumption of soy (the major dietary source of phytoestrogens) was correlated with reduced breast cancer risk in a Singapore population.

Adlercreutz and co-workers[37,38] have suggested that lignans and isoflavanoids stimulate the synthesis of SHBG in the liver, supported by studies showing a positive correlation between urinary excretion of lignans and phytoestrogens and plasma concentration of SHBG in Finnish omnivorous and vegetarian females.[38] Elevated SHBG causes a decrease in the percent of free estradiol, thereby lowering estrogen exposure, another mechanism by which phytocompounds could protect against breast cancer.

VII. Hypotheses on the Effect of Dietary Fat and Fiber on Estrogen Metabolism

Endogenous estrogenic hormones are excreted in bile as glucuronide, sulfate or glutathione conjugates. In the bowel lumen these conjugated estrogens are deconjugated by intestinal bacterial enzymes, thereby permitting reabsorption across the bowel mucosa into the bloodstream. Studies in humans and animals have shown that a high intake of dietary fat results in a significant increase in fecal bacterial beta-glucuronidase activity (Figure 1).[23,40] Therefore, it follows that individuals eating a high fat diet would have increased estrogen deconjugation with a resultant increase in intestinal absorption of estrogens, leading to higher plasma estrogen concentrations and lower fecal excretion of estrogen. Eating a low fat diet reduces intestinal reabsorption, increases fecal excretion and lowers plasma estrogen levels. In addition, a high fiber diet (Figure 2) can trap estrogens in the intestine and reduce bacterial beta-glucuronidase activity. Either mechanism causes decreased reabsorption of estrogens from the intestine, resulting in higher fecal excretion and lower concentrations of estrogens in the plasma. In sum, a low fat, high fiber diet is associated with low plasma estrogen levels.

VIII. Estrogen Levels in Vegetarian and Omnivore Women

Table 1 presents the demographic and dietary data for 10 vegetarian and 10 nonvegetarian premenopausal American women studied in our laboratory.[23] Blood, urine, and feces, were collected on four occasions, at intervals of approximately four months, during the mid-follicular phase of the menstrual cycle. The vegetarians had significantly higher fecal excretion of all three estrogens measured (Table 2). Based on total 24 hour estrogen output in the feces, the vegetarians excreted 2 to 3 fold higher amounts than the omnivores.

In Table 3 the amounts of several important estrogens excreted in urine over a 24 hour period are shown for vegetarians and omnivores. The omnivores excreted significantly higher estriol (52%) and higher estriol-3-glucuronide (38%). Estriol undergoes the most extensive enterohepatic circulation of the various estrogens, and it is a good marker for the extent of this process. Estriol-3-glucuronide is formed in the intestinal mucosal cell after absorption of estriol from the lumen; changes in its concentration are a reflection of intestinal reabsorption.[41] These findings suggest that omnivores absorb more estriol into the bloodstream and lose less estriol in the feces than vegetarians.

Plasma estrone and estradiol concentrations are also different in vegetarian and omnivore women (Table 4). The omnivores had 17% higher estrone and 23% higher estradiol, but the differences did not achieve statistical significance. There was, however, a significant

High Fat Intake

Increases bacterial deconjugation

Increases intestinal reabsorption
of steroid hormones

Increases enterohepatic circulation

Increases plasma hormone levels
Increases urinary excretion
Decreases fecal excretion

Figure 1. Hypothesis relating high dietary fat intake to estrogen status.

High Fiber Intake

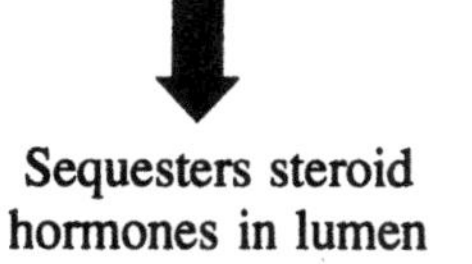

Sequesters steroid
hormones in lumen

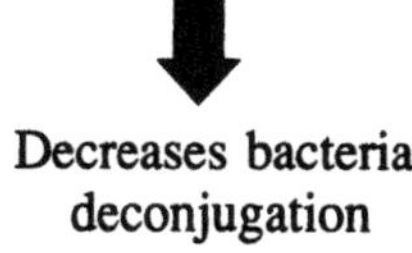

Decreases bacterial
deconjugation

Decreases intestinal reabsorption
of steroid hormones

Decreases enterohepatic circulation

Decreases plasma hormone levels
Decreases urinary excretion
Increases fecal excretion

Figure 2. Hypothesis relating high dietary fiber intake to estrogen status.

Table 1. Demographic and Dietary Data on 10 Omnivorous and 10 Vegetarian Women

	Omnivores		Vegetarians	
Demographic Data (arithmetic mean±S.E.)				
Age (yr)	25.5 ± 1.0		26.5 ± 0.9	
Height (cm)	166.8 ± 2.7		163.7 ± 2.4	
Weight (kg)	63.8 ± 2.6		59.9 ± 2.9	
Body-mass index: weight (kg)/height (cm)^2	22.7 ± 0.5		22.6 ± 1.0	
Dietary Data (daily consumption)[a]				
Kilocalories	1572	(1496-1652)	1649	(1625-1734)
Protein (g)	62	(59-65)	55	(52-58)
Animal protein (%)	43[b]		7	
Dietary fiber (g)	12[b]	(11-13)	28	(25-33)
Dietary fat				
Total fat	70	(63-74)	56	(53-60)
Total fat (g/1000 cal)	44[c]	(42-46)	34	(31-37)
Saturated fat (g)	27[d]	(25-29)	17	(16-19)
Saturated fat (g/1000 cal)	17[c]	(16-18)	10	(9-12)
Calories as fat (%)	40		30	

[a] With the exception of total fat and calories from fat, all dietary data are expressed as geometric means, with the range of S.E. in parentheses.

[b] $p<0.001$.

[c] $p<0.05$.

[d] $p<0.07$.

Table 2. Fecal Excretion of Estrogens in 10 Omnivorous and 10 Vegetarian Women[a]

Hormone	Omnivores		Vegetarians	
	nmol/24 hours			
Estrone	0.83	(0.70-0.99)	1.96	(1.68-2.29)[c]
Estradiol	0.61	(0.52-0.72)	1.52	(1.30-1.78)[b]
Estriol	0.72	(0.63-0.83)	1.72	(1.50-1.98)[c]
Total estrogen (estrone, estradiol, estriol)	2.33	(2.01-2.70)	5.40	(4.70-6.21)[b]

[a] Values are presented as geometric means with S.E. ranges in parentheses. Calculations are based on four collections per subject.

[b] $p<0.03$.

[c] $p<0.01$.

Table 3. Urinary Excretion of Estrogens in 10 Omnivorous and 10 Vegetarian Women[a]

Hormone	Omnivores		Vegetarians	
	nmol/24 hours			
Estrone	15.3	(13.8-17.0)	16.8	(15.0-18.8)
Estradiol	9.3	(8.6-10.0)	9.3	(8.5-10.2)
Estriol	21.0	(19.1-23.1)	13.2	(11.6-15.0)[b]
Estriol-3-glucuronide	38.5	(36.0-41.2)	28.0	(24.8-31.6)

[a] Values are presented as geometric means with S.E. ranges in parentheses. Calculations are based on four collections per subject.

[b] p<0.05.

Table 4. Plasma Estrogen and Androgen Levels in 10 Omnivorous and 10 Vegetarian Women[a]

Hormone	Omnivores		Vegetarians	
	nmol/liter			
Estrone	0.40	(0.36-0.44)	0.34	(0.32-0.36)
Estradiol	0.32	(0.29-0.36)	0.26	(0.23-0.29)

[a] Values are presented as geometric means with S.E. ranges in parentheses. Calculations are based on four collections per subject.

negative correlation between plasma estrogen concentration and fecal estrogen excretion (p<0.003). These data suggest a decreased enterohepatic circulation as reflected in higher fecal estrogen excretion and a lower concentration of circulating estrogen in vegetarians.

IX. Estrogen Levels in Vegans and Oriental Women

A subsequent study included pre- and postmenopausal Caucasian omnivores, lacto-ovovegetarians, vegans and recent Asian immigrants to Hawaii (primarily Vietnamese),[23,24] all with distinctly different dietary intakes. The Boston vegetarians were primarily lacto-ovovegetarians, which meant that they consumed dairy products and eggs with a fat intake of 30% of calories and 28 gms of dietary fiber per day. The vegans, in contrast, did not eat any animal products, and their intake was 20-22% of calories as fat, compared to 40% in omnivores and 30% in lacto-ovovegetarians. The vegans also ate a higher fiber diet with an average intake of approximately 30 grams per day. The Oriental women consumed 19-22% of calories as fat and 16-20 gms of dietary fiber per day.

Among premenopausal women, the omnivores had higher serum estrone (28%) and estradiol (24%) than vegans (p<0.05) (Table 5). Omnivores also had nonstatistically significant higher estrone (14%) compared to lacto-ovovegetarians. The postmenopausal omnivore women had higher serum estrone (43%) and estradiol (62%) concentrations compared to those levels in postmenopausal vegans. These changes were not significant due

Table 5. Serum Estrogen Concentrations in Omnivores, Lacto-ovovegetarians, and Vegetarians

Group	Number of Determinations	Estrone	Estradiol
		nmol/liter	
Premenopausal			
Omnivore	20	.322 ± .096[a]	.251 ± .060[a]
Lacto-ovovegetarian	12	.282 ± .086	.255 ± .088
Vegan	15	.251 ± .087[b]	.203 ± .062[b]
Postmenopausal			
Omnivore	16	.162 ± .062	.155 ± .040
Lacto-ovovegetarian	4	.155 ± .015	.111 ± .017
Vegan	5	.113 ± .045	.071 ± .052

± standard error
[a,b] a:b p<0.05.

to the small sample size (5 vegan postmenopausal women); however, the magnitude of these changes was large, suggesting that postmenopausal vegans have lower circulating estrogen.

As in previous studies with lacto-ovovegetarians, fecal estrogen excretion and serum values were different in Caucasian omnivores and Oriental women. The Oriental premenopausal women excreted 2-3 times more total estrogen in the feces, as well as higher amounts of the individual estrogens, estrone, estradiol and estriol compared to omnivore women (Table 6). The premenopausal Oriental women had significantly lower plasma concentrations of estrone and estradiol compared to omnivores (Table 7); the postmenopausal Oriental women had significantly lower estradiol plasma levels, but estrone levels were similar to those in omnivores.

The association between daily dietary intake of fat and fiber and plasma estrogen concentration in this study was analyzed.[22] Total dietary fat intake, expressed as g/24 hr, was positively correlated with plasma concentrations of estradiol (r=0.65, p=0.001) and estrone (r=0.57, p=0.005). A positive correlation was also observed between saturated fat intake and plasma estrone concentration (r=0.52, p=0.012). No correlation was seen, however, between polyunsaturated fat consumption and plasma estrone or estradiol concentrations. When intake of dietary fiber was analyzed, an inverse correlation was found with plasma levels of estrone (r=-0.55, p=0.008) and estradiol (r=-0.65, p=0.001).

X. Effects of Diet in a Controlled, Metabolic Setting

Recent studies from our laboratory have focused on feeding defined diets to Caucasian women in the Boston area.[24,42] All meals were prepared in a tightly controlled metabolic kitchen, and with the exception of lunch and dinner on Saturdays and the three meals on Sunday, were all eaten in a metabolic unit. Several dietary constituents were varied, including fat, fiber, the relative amounts of saturated, monosaturated and polyunsaturated fats (S:M:P ratio) and dietary cholesterol. Forty-eight premenopausal omnivore women participated in fifty-eight study protocols.[24,42] In the first series they were shifted from a typical omnivore diet (40% fat, 12 gms of fiber) which they had eaten for 1 month to a low fat

Table 6. Fecal Excretion of Estrogens in Caucasian American and Oriental Immigrant Women[a]

| | Premenopausal | | |
Hormones[a]	Caucasians (n=10)	Orientals (n=12)	p
Estrone	1.10 (0.82-1.46)	2.06 (1.68-2.51)	<0.05
Estradiol	0.54 (0.40-0.69)	1.68 (1.36-2.06)	<0.02
Estriol	0.93 (0.66-1.30)	3.50 (2.95-4.13)	<0.02
Total estrogen	2.64 (1.95-3.56)	7.53 (6.38-8.88)	<0.02

[a] Values are presented as geometric means of nmol/24 hour with S.E. ranges in parentheses.

Table 7. Plasma Estrogen Levels in Pre- and Postmenopausal Caucasian American and Oriental Immigrant Women[a]

| | Premenopausal | | Postmenopausal | |
Hormones[a]	Caucasians (n=10)	Orientals (n=12)	Caucasians (n=10)	Orientals (n=9)
Estrone	0.32 (0.30-0.34)	0.24 (0.22-0.25)[b]	0.14 (0.13-0.15)	0.13 (0.10-0.15)
Estradiol	0.25 (0.24-0.26)	0.14 (0.12-0.15)[c]	0.10 (0.09-0.11)	0.03 (0.02-0.04)[c]
Estrone plus estradiol	0.57 (0.54-0.60)	0.39 (0.36-0.42)[c]	0.24 (0.22-0.26)	0.17 (0.14-0.21)[d]

[a] Values are presented as geometric means of nmol/L with S.E. ranges in parentheses.
[b] p<0.02.
[c] p<0.001.
[d] p=0.05.

(20-25%), high fiber (40 gms) diet (Table 8). All data were adjusted for age and body mass index (BMI). For each of the hormones and for the sex hormone-binding globulin (SHBG), the shift to a low fat, high fiber diet resulted in a decrease in serum concentration. The largest decrease (30%) was for the major storage estrogen, estrone sulfate.

XI. Conclusions

A number of studies on the effects of diet on sex hormone levels support the hypothesis that diet influences breast cancer risk by altering the endogenous hormonal milieu. Increased intake of fiber and decreased intake of fat act independently to lower sex hormone levels.

Prentice et al.[43] concluded that a 17% reduction of plasma estradiol may in large part explain international differences in breast cancer incidence. Both in the studies of "free-living" populations and in the defined diet studies conducted in our laboratory, a 10-20% reduction in plasma estrogen levels, and in the case of estrone sulfate, a 30% reduction, were noted in women consuming less then 30% of calories from fat and more then 20 grams of fiber per day. Therefore, it appears that lowering fat and increasing fiber in the standard diet

Table 8. The Effects of a Low Fat–High Fiber Diet on Serum Hormone Concentrations

Hormone	% Change[a]	P value
Estrone	-9.2	0.04[b]
Estrone sulfate	-30.3	<0.001[b]
Estradiol	-10.5	0.06
Free estradiol	-10.5	0.09
Testosterone	-12.4	0.04[b]
Androstenedione	-9.6	0.01[b]
SHBG	-14.7	0.001[b]

[a] A negative sign indicates that the concentration of the hormone decreased after consuming a low fat–high fiber diet. n=48 subjects; 207-226 individual measurements per hormone.

[b] Statistically significant.

of Western women results in reduced estrogen exposure and could lower the risk of developing breast cancer.

Acknowledgments

Supported by NIH grants R01 CA54349 and R37 CA45128.

References

1. Gray, G.E., M.C. Pike, and B.E. Henderson, Breast-cancer incidence and mortality rates in different countries in relation to known risk factors and dietary practices, *Br. J. Cancer* 39:1 (1979).
2. Hems, G., Epidemiological characteristics of breast cancer in middle and late age, *Br. J. Cancer* 24:226 (1970).
3. Armstrong, B. and R. Doll, Environmental factors and cancer incidence and mortality in different countries with special reference to dietary practices, *Int. J. Cancer* 15:617 (1975).
4. Carroll, K.K., E.B. Gammal, and E.R. Plunkett, Dietary fat and mammary cancer, *Can. Med. Assoc. J.* 98:590 (1968).
5. Lea, A.J., Dietary factors associated with death-rates from certain neoplasms in man, *Lancet* ii:332 (1966).
6. Wynder, E.L., Identification of women at high risk for breast cancer, *Cancer* 24:1235 (1969).
7. Goodwin, P.J. and N.F. Boyd, Critical appraisal of the evidence that dietary fat intake is related to breast cancer risk in humans, *J. Natl. Cancer Inst.* 79:473 (1987).
8. Ingram, D.M., Trends in diet and breast cancer mortality in England and Wales 1928-1977, *Nutr. Cancer* 3:75 (1981).
9. Dunn, J.E., Breast cancer among American Japanese in the San Francisco Bay area, *Monogr. Natl. Cancer Inst.* 47:157 (1977).
10. Willett, W.C., M.J. Stampfer, G.A. Colditz, B.A. Rosner, C.H. Hennekens, and F.E. Speizer, Dietary fat and the risk of breast cancer, *N. Engl. J. Med.* 316:22 (1987).
11. Willett, W.C., D.J. Hunter, M.J. Stampfer, G. Colditz, J.E. Manson, D. Spiegelman, B. Rosner, C.H. Hennekens, and F.E. Speizer, Dietary fat and fiber in relation to risk of breast cancer. An 8-year follow-up, *JAMA* 268:2037 (1992).

12. Phillips, R.L. and D.A. Snowden, Association of meat and coffee use with cancers of the large bowel, breast and prostate among Seventh-Day Adventists: Preliminary results, *Cancer Res.* 43 (Suppl):2403s (1983).

13. Hirayama, T., Epidemiology of breast cancer with special reference to the role of diet, *Prev. Med.* 7:173 (1978).

14. Phillips, R.L., Role of life-style and dietary habits in risk of cancer among Seventh-Day Adventists, *Cancer Res.* 35:3513 (1975).

15. Miller, A.B., A. Kelly, N.W. Choi, V. Matthews, R.W. Morgan, L. Munan, J.D. Burch, J. Feather, G.R. Howe, and M. Jain, A study of diet and breast cancer, *Am. J. Epidemiol.* 107:499 (1978).

16. Howe, G.R., The use of polytomous dual response data to increase power in case-control studies: an association between dietary fat and breast cancer, *J. Chronic Diseases* 38:663 (1985).

17. Lubin, J.H., P.E. Burns, W.J. Blot, R.G. Ziegler, A.W. Lees, and J.F. Fraumeni Jr., Dietary factors and breast cancer risk, *Int. J. Cancer* 28:685 (1981).

18. Lee, H.P., L. Gourley, S.W. Duffy, J. Esteve, J. Lee, and N.E. Day, Risk factors for breast cancer by age and menopausal status: a case-control study in Singapore, *Cancer Causes Control* 3:313 (1992).

19. Howe, G.R., T. Hirohata, T.G. Hislop, K. Katsouyanni, F. Lubin, E. Marubini, B. Modan, T. Rohan, P. Toniolo, and Y. Shunzhang, Dietary factors and risk of breast cancer: combined analysis of 12 case-control studies, *J. Natl. Cancer Inst.* 82:561 (1990).

20. Kelsey, J.L., A review of the epidemiology of breast cancer, *Epidemiol Rev.* 1:74 (1979).

21. Russo, J. and I.H. Russo, Biology of breast disease, *Lab. Invest.* 57:112 (1987).

22. Armstrong, B.K., J.B. Brown, and H.T. Clarke, Diet and reproductive hormones: a study of vegetarian and non-vegetarian postmenopausal women, *J. Natl. Cancer Inst.* 67:761 (1981).

23. Goldin, B.R., H. Adlercreutz, S.L. Gorbach, J.H. Warram, J.T. Dwyer, L. Swenson, and M.N. Woods, Estrogen excretion patterns and plasma levels in vegetarians and omnivorous women, *N. Engl. J. Med.* 307:1542 (1982).

24. Goldin, B.R., H. Adlercreutz, S.L. Gorbach, M.N. Woods, J.T. Dwyer, T. Conlon, E. Bohn, and S.N. Gershoff, The relationship between estrogen levels and diets of Caucasian American and Oriental immigrant women, *Am. J. Clin. Nutr.* 44:945 (1986).

25. Woods, M.N., S.L. Gorbach, C. Longcope, B. Goldin, J.T. Dwyer, and A. Morrill-LaBrode, Low-fat, high-fiber diet and serum estrone sulfate in premenopausal women, *Am. J. Clin. Nutr.* 49:1179 (1989).

26. Schultz, T.D. and J.E. Leklem, Nutrient intake and hormonal status of premenopausal vegetarian Seventh-Day Adventists and premenopausal nonvegetarians, *Nutr. Cancer* 4:247 (1983).

27. Rose, D.P., M. Goldman, J.M. Connolly, and L.E. Strong, High-fiber diet reduces serum estrogen concentrations in premenopausal women, *Am. J. Clin. Nutr.* 54:520 (1991).

28. Farnsworth, N.R., A.S. Bingel, G.A. Cordell, F.A. Crane, and H.H.S. Fong, Potential value of plants as sources of new antifertility agents II, *J. Pharm. Sci.* 64:717 (1975).

29. Bennetts, H.W., E.J. Underwood, and F.L. Shier, A specific problem of sheep on subterranean clover pastures in Western Australia, *Aust. J. Agric. Res.* 22:131 (1946).

30. Setchell, K.D.R. and H. Adlercreutz, Mammalian lignans and phytoestrogens: recent studies on their formation, metabolism and biological role in health and disease, *in: Role of Gut Flora in Toxicity and Cancer*, I.R. Rowland, ed., Academic Press, New York (1988).

31. Setchell, K.D.R., A.M. Lawson, F.L. Mitchell, H. Adlercreutz, D.N. Kirk, and M. Axelson, Lignans in man and animal species, *Nature* 287:740 (1980).

32. Adlercreutz, H., T. Fotsis, C. Bannwart, K. Wahala, T. Makela, G. Brunow, and T. Hase, Determination of urinary lignans and phytoestrogen metabolites, potential antiestrogens and anticarcinogens, in urine of women on various habitual diets, *J. Steroid Biochem.* 25:791 (1986).

33. Mousavi, Y. and H. Adlercreutz, Enterolactone and estradiol inhibit each other's proliferative effect on MCF-7 breast cancer cells in culture, *J. Steroid Biochem. Mol. Biol.* 41:615 (1992).

34. Peterson, G. and S. Barnes, Genistein and biochanin A inhibit the growth of human prostate cancer cells but not epidermal growth factor receptor tyrosine autophosphorylation, *Prostate* 22:335 (1993).

35. Adlercreutz, H., T. Fotsis, and J. Lampe, Quantitative determination of lignans and isoflavonoids in plasma of omnivorous and vegetarian women by isotope dilution gas chromatography — mass spectrometry, *Scand. J. Clin. Lab. Invest.* 53 (Suppl 215):5 (1993).

36. Mousavi, Y. and H. Adlercreutz, Genistein is an effective stimulator of sex hormone-binding globulin production in hepatocarcinoma human liver cancer cells and suppresses proliferation of these cells in culture, *Steroids* 58:301 (1993).

37. Adlercreutz, H., Western diet and Western diseases: some hormonal and biochemical mechanisms and associations, *Scand. J. Clin. Lab. Invest.* 50 (Suppl 201):3 (1990).

38. Adlercreutz, H., K. Hockerstedt, C. Bannwart, E. Hamalainen, T. Fotsis, and S. Bloigu, Association between dietary fiber, urinary excretion of lignans and isoflavonic phytoestrogens and plasma non-protein bound sex hormones in relation to breast cancer, *Prog. Cancer Res. Ther.* 35:409 (1988).

39. Adlercreutz, H., H. Honjo, A. Higashi, T. Fotsis, E. Hamalainen, T. Hasegawa, and H. Okada, Urinary excretion of lignans and isoflavonoid phytoestrogens in Japanese men and women consuming a traditional Japanese diet, *J. Clin. Nutr.* 54:1093 (1991).

40. Goldin, B.R., L. Swenson, J.T. Dwyer, M. Sexton, and S.L. Gorbach, Effect of diet and *Lactobacillus acidophilus* supplements on human fecal bacterial enzymes, *J. Natl. Cancer Inst.* 64:255 (1980).

41. Adlercreutz, H. and T. Luukkainen, Biochemical and clinical aspects of the enterohepatic circulation of oestrogen, *Acta. Endocrinol.* 124 (Suppl):101 (1967).

42. Goldin, B.R., M.N. Woods, D.L. Spiegelman, C. Longcope, A. Morrill-LaBrode, J.T. Dwyer, L.J. Gaultiere, E. Hertzmark, and S.L. Gorbach, The effect of dietary fat and fiber on serum estrogen concentrations in premenopausal women under controlled dietary conditions, *Cancer* (Suppl) (In Press).

43. Prentice, R., D. Thompson, C. Clifford, S. Gorbach, B. Goldin, and D. Byar, Dietary fat reduction and plasma estradiol concentration in healthy postmenopausal women, *J. Natl. Cancer Inst.* 82:129 (1990).

Chapter 5

Dietary Fat Effects on Animal Models of Breast Cancer

WILLIAM T. CAVE, JR.

I. Introduction

Carcinogenesis is a multistage process which is frequently influenced by environmental factors. Elements in our nutritional environment have been repeatedly identified as important influences in the final expression of many human cancers. In this regard, both epidemiological studies and laboratory investigations have indicated that high fat diets enhance the growth and development of breast cancer. The biochemical mechanisms responsible for this dietary effect remain obscure at present, but the potential benefits of better understanding this important relationship are well recognized. Since epidemiology has a limited capability for defining such complex biochemical events, animal tumors have become the main experimental models for study.

II. Lipid Effects on Mammary Tumorigenesis

The ability of dietary fat to enhance mammary tumorigenesis in rodents has been known for almost half a century. Tannenbaum,[1-4] using strains of mice with high incidences of spontaneous mammary carcinomas, reported that it was possible to increase the rate of formation of mammary carcinomas by varying the concentration of dietary fat from 2-26% in equicaloric diets. In 1956 Davis et al.[5] reported a similar effect of dietary fat on mammary tumors in rats. They noted that normal female Sprague-Dawley rats allowed to live out their lifespans on standard laboratory chow had a benign and malignant spontaneous mammary tumor incidence of 57%, while animals on a special fat-rich diet had an incidence of 80%. Carroll and Khor subsequently observed that DMBA-treated rats fed diets containing 10% and 20% corn oil (w/w) had more mammary adenocarcinomas and a shorter tumor latency than animals fed 0.5% and 5% corn oil.[6] Since then, numerous laboratories have confirmed or extended these initial observations using a variety of rodent mammary tumor models: spontaneous tumors, transplanted tumors, carcinogen-induced tumors (dimethylbenz-

---

William T. Cave, Jr. ● Endocrine Unit, Department of Internal Medicine, University of Rochester Medical Center, Rochester, New York

Diet and Breast Cancer, Edited under the auspices of the
American Institute for Cancer Research, Plenum Press, New York, 1994

[a]anthracene [DMBA], N-methyl-N-nitrosourea [NMU], 2-acetylaminofluorene), diethylstilbestrol (DES)-induced tumors, and X-ray induced tumors.[7] These results, together with the lack of evidence that variations in other factors such as proteins, vitamins or minerals above the levels required for normal maintenance have any similar influence on the genesis or growth of mammary tumors,[8] have led most investigators to conclude that dietary fat has a unique mammary tumor promoting activity. Additional confirmation of this conclusion has been provided recently by the important statistical review of data extracted from over 100 animal experiments by Freedman *et al.*[9] They concluded that there was a specific enhancing effect of dietary fat on mammary tumor development that was independent of increased caloric intake.

The animal models, in agreement with the human epidemiological observations, also indicated that this effect of dietary fat on mammary tumor development was not universally shared by all forms of neoplasia. Engle and Copeland,[10] for example, reported that while they could initiate tumors in several organs (mammary gland, ear-duct, and liver) of female rats with 2-acetylaminofluorene, only the mammary tumors grew when the animals were placed on a high fat diet. To date, it has been shown that this sensitivity of mammary tumor cells to dietary lipid is shared, to some extent, by melanomas,[11] skin,[12] liver,[13] colon[14] and pancreatic tumors,[15] but not by sarcomas,[16] primary lung adenomas,[1] and submaxillary gland tumors,[12] or spontaneous and induced leukemia.[17]

Once it was appreciated that quantitative differences in dietary fat affected mammary tumor development, subsequent investigations sought to evaluate the importance of specific qualitative differences in the composition of dietary fat. In this regard, Carroll and Hopkins[18] using the DMBA tumor model, discovered that there was a significant difference between the tumor enhancing effects of the n-6 polyunsaturated fats and saturated fats. They reported that polyunsaturated fats (e.g. cottonseed oil, sunflower oil, and corn oil), when fed as 20%(w/w) of the diet, induced nearly twice as many mammary tumors in rats as did equivalent levels of saturated fats (e.g. coconut oil, butter or tallow). Furthermore they noted that 20% saturated fat diets produced little or no increase in the tumor yield over that obtained with a low n-6 polyunsaturated fat (0.5% corn oil) diet. This has been confirmed by other investigators.[7] Moreover studies with primary cell cultures of DMBA-induced rat mammary carcinomas have shown that unsaturated free fatty acids (e.g. oleic acid (C 18:1) or linoleic acid (C 18:2)) added to the culture media enhance the growth rate of mammary tumor cells, while saturated free fatty acids (e.g. stearic acid (C 18:0)) inhibit their growth.[19,20] There has been no evidence of any direct correlation between the tumor enhancing effect of n-6 polyunsaturated fats, however, and their absolute degree of unsaturation. In fact, the ratio of saturated to unsaturated fatty acids in the diet can be as high as 6:1 and still provide the same tumor yield as diets in which the ratio is 1:10.[21]

Hopkins and Carroll[22] have clarified this complex interrelationship between the amount of polyunsaturated fat in the diet and mammary tumor development somewhat by observing that, when DMBA treated rats were fed a small amount of n-6 polyunsaturated fat (e.g. 3%) along with 17% saturated fat, their tumor development was equivalent to that of rats fed a 20% n-6 polyunsaturated fat diet. They concluded that there must be a "basal" requirement for polyunsaturated fat in mammary tumorigenesis that cannot be adequately satisfied by saturated fats alone but which can be met by supplements as little as 3% of n-6 polyunsaturated fat. Since ethyl linoleate alone, when combined with high saturated fat diets, can meet this requirement, it has been inferred that this essential n-6 polyunsaturated fatty acid is the critical component required for this effect. It has been noted, however, that such small amounts of dietary n-6 polyunsaturated fats, by themselves, are not associated with a high tumor yield. Thus maximal tumor growth occurs only when both a high total fat level and a "basal" amount of n-6 polyunsaturated are present in the diet. Many others have corroborated these results, although the minimum amount of essential fatty acid required for optimal tumor development may vary somewhat among different mammary tumor model systems.[23-25]

The tumor promoting capabilities of dietary fats containing high levels of monoenoic fatty acids (e.g. olive oil) have also been evaluated. While there is no complete unanimity of opinion about the results of these studies, the findings do seem to indicate that these monoenoic fats do not have the tumor promoting properties of the n-6 polyunsaturated fats.[26,27] Thus, like the medium chain length fatty acids to some degree, they behave in this regard as saturated fats.[28] Similar to the saturated fats, Lasekan[29] found that while DMBA treated rats fed a 20% olive oil diet alone had a reduced tumor incidence, the addition of 2.3% linoleate to the olive oil overcame these effects. Recently, Buckman *et al.*[30] have also observed from their 4526 metastatic mammary tumor model that oleate does not further enhance metastasis in mice fed a high linoleate diet, but it does significantly increase metastasis in mice fed the low (3%) linoleate diets.

The experimental evidence also indicates that there are differences in tumor promoting capability among the different families of polyunsaturated fatty acids. In this regard, Jurkowski and Cave[31] have reported that diets containing high levels of omega-3 polyunsaturated fatty acids do not promote mammary tumor development in the NMU tumor model in the same manner as diets containing equivalent levels of omega-6 polyunsaturated fatty acids. They examined the effect of diets containing increasing amounts (0.5%, 3%, or 20%) of menhaden oil (high omega-3 polyunsaturated fatty acid content) on tumor development and then compared the tumor promoting effects of diets containing various levels of menhaden oil with those containing equivalent levels of corn oil (high omega-6 polyunsaturated fatty acid content). As the percentage of menhaden oil increased, there was a progressive lengthening of the tumor latent period, as well as a reduction in tumor incidence and tumor burden. In sharp contrast, the rats on the 20% corn oil diet had a high tumor incidence and the shortest latent period. Braden and Carroll[32] have also reported finding similar differences in the DMBA mammary tumor model. They examined six diet groups: 20% corn oil, 10% corn oil, 3% corn oil, 20% menhaden oil, 10% menhaden oil, and 3% menhaden oil, and observed that there was an increase in the tumor yield and a decrease in the tumor latent period in the groups receiving the higher levels of corn oil, relative to the groups receiving the equivalent levels of menhaden oil. Whether this tumor inhibiting characteristic of the longer chain omega-3 polyunsaturated fatty acids typically found in marine oils is also shared by the shorter chain omega-3 fatty acids such as linolenic acid (C18:3) in linseed oil is less clear. The mean time of tumor latency for spontaneous mammary tumors in C3H mice has been reported to be 5-8 weeks longer in mice fed a 10% linseed oil diet than in those fed a 10% corn oil or 10% safflower oil diet.[33] However, *in vitro* studies with cultures of dimethylbenz[a]anthracene-induced mammary tumor cells indicate that very low media concentrations of linolenic acid are stimulatory, while higher concentrations are inhibitory.[20]

More recent investigations have sought to compare the tumor enhancing characteristics of certain dietary blends of omega-3 and omega-6 fatty acids. Cave *et al.*[34] for example, have compared the effects of the following diets: 20% corn oil, 15% corn oil + 5% menhaden oil, 15% menhaden oil + 5% corn oil, 20% menhaden oil in the NMU tumor model. They observed that the animals on the 20% menhaden oil diet had the longest tumor latent period, while the animals on the 20% corn oil and 15% corn oil + 5% menhaden oil had the shortest latent periods. The latent period of the 15% menhaden oil + 5% corn oil group, however, was considerably longer than that of the 20% corn oil group, suggesting that the high proportion of menhaden oil had some ability to delay tumorigenesis, even in the presence of 5% corn oil. Similarly, using the DMBA model, Ip *et al.*[35] reported that rats fed 12% menhaden oil + 8% corn oil have a lower incidence of mammary tumor than animals fed 20% corn oil, while Karmali *et al.*[36] noted decreased mammary tumor development in animals receiving 8% corn oil + 15.5% fish oil or 3% corn oil + 20.5% fish oil.

Both non-metastatic and metastatic mammary tumor transplant models have also provided experimental evidence of the contrasting tumor promoting capabilities of high omega-3 and omega-6 polyunsaturated fat diets. In one investigation Karmali *et al.*[37] reported that high doses of fish oil reduced the growth of subcutaneously transplanted R3230AC mammary adenocarcinoma in female F344 rats. Subsequently, Gabor and Abraham[38] observed that high dietary levels of either menhaden oil or hydrogenated cottonseed oil reduce the development of a transplantable BALB/c mammary adenocarcinoma in mice. They further noted that the tumor enhancing effects of the corn oil were significantly inhibited when the menhaden oil:corn oil ratio was 9:1. Kort *et al.*[39] have also found that a 25% fish oil diet inhibits the development of the BN472 transplantable mammary adenocarcinoma in female BN/Bi rats relative to one containing 25% cacao butter, but they noted that the tumor inhibition was more pronounced when the fish oil diets were initiated before transplantation than when given only after transplantation. Using human breast carcinoma cells (MCF-7) inoculated into athymic (nude) MF1 mice, Pritchard *et al.*[40] have shown that a 10% fish oil diet can inhibit tumor development. Blank and Ceriani[41,42] and Borgeson *et al.*[43] have indicated that human breast carcinoma (MX-1) cells inoculated into athymic BALB/c nude mice are quite sensitive to the inhibiting effects of an 11% fish oil diet. Furthermore, they also reported that the fish oil diets are synergistic with I-131 labelled monoclonal antibodies in reducing tumorigenesis and can enhance tumor responsivity to chemotherapy with mitomycin C and doxorubicin.

III. Biochemical Mechanisms

When viewed either through the perspective of the classical two-step model of carcinogenesis of Berenblum,[44] or by more recent interpretations of this model,[45] the primary action of lipids in tumor development has been consistently perceived as being more critically involved in the process of tumor promotion than in the process of tumor initiation. More specifically, dietary lipids do not appear to generate carcinogenic chromosomal mutations themselves, but rather seem to modulate the function of already mutated cells. This distinctive effect of lipids on tumor promotion was initially illustrated through a cross over experiment, which demonstrated that dietary fat enhanced tumor growth more effectively when given after chemical carcinogen administration, than when given only prior to, or during carcinogen administration.[46] *In vitro* experiments using cell cultures have also confirmed the promotional activity of lipids. A variety of tumor transplant studies have shown, quite convincingly, that it is the dietary lipid environment of the recipient, rather than that of the donor, that is responsible for the ultimate pattern of tumor development.[47-50]

There appears to be no specific time limit following tumor initiation during which the cells lose their responsiveness to dietary lipid. Even when the high fat diet has been delayed for as long as 20 weeks after carcinogen administration,[51] the tumor yield can still be enhanced. Similarly, rats fed a high fat diet for equal lengths of time (3 weeks), but at different periods during early tumorigenesis (0, 2, or 4 weeks after carcinogen administration) acquire equivalent increases in mammary tumor development.[52] Thus, the stimulatory effects of high fat diets on mammary tumor growth do not seem to be permanently induced, but rather they require continuous dietary treatment. Some investigators, in fact, have proposed that there is a dose-response relationship, in which the extent of mammary tumor development is dependent on the duration of the high fat diet.[24]

Despite the rather extensive amount of experimental data documenting this effect of dietary lipids on the promotional phase of mammary tumor development, relatively little is known about the specific biochemical processes involved. Both indirect effects on the host tissue environment, and/or direct effects on the internal environment in the tumor cells may be involved. One area of investigation that has received considerable attention has been the effect of dietary fat on the host's hormonal environment. The evidence indicates that normal rodent mammary gland development involves the interactions of a number of hormones (e.g. prolactin, growth hormone, estrogens, progesterone, glucocorticoids, insulin, and thyroxine), but deprivation and replacement studies indicate that prolactin and estradiol are the critical regulators of experimental mammary tumorigenesis.[53,54] As such, it has been repeatedly shown in hormone-sensitive carcinogen-induced mammary tumor models that the tumor promoting effects of dietary lipids are virtually absent in an endocrinologically deficient host. For example, the administration of the prolactin inhibiting drug, bromocriptine,[55] or ovariectomy performed after carcinogen administration, have been shown to inhibit completely mammary tumor growth in animals on either a high or low fat diet.[56] Furthermore, if such carcinogen treated, ovariectomized rats are subsequently given estradiol replacement and haloperidol (a prolactin stimulating drug), the animals on the high fat diet will again express enhanced tumor development. Likewise, *in vitro* experiments with primary cultures of DMBA-induced mammary tumors also indicate that maximal mammary tumor cell growth requires the presence of both polyunsaturated fatty acids and hormone in the culture media.[20,21]

Precisely how dietary lipids interact with these hormones to promote mammary tumor development is unresolved. Although it was postulated initially that dietary fat might enhance the production or secretion of certain mammotrophic hormones, serum samples obtained from indwelling right atrial catheters in conscious unstressed rats, have indicated that the prolactin, progesterone, and estradiol levels of animals on high fat diets are qualitatively and quantitatively identical to those of normal rats throughout the estrous cycle.[57] Furthermore, the pituitary glands of the high fat diet rats show no evidence *in vitro* of any difference in the synthesis or secretion of prolactin and growth hormone.[58] This lack of evidence of any changes inducible by dietary lipid in the levels of circulating hormones, therefore, suggests that either dietary lipids must alter the sensitivity of mammary tumor cells to their normal hormonal environment, or that they can promote mammary tumorigenesis only in the presence of an adequate hormonal milieu.

There is a considerable amount of evidence to indicate that qualitative and quantitative differences in dietary lipid can significantly alter the membrane fatty acid compositions of both normal and neoplastic cells.[23,31] As such, it has been proposed that dietary lipids may influence tumor development by modifying the physical-chemical environment of hormone receptors and/or enzymes of the tumor cells. It is well documented that many polypeptide molecules (e.g. prolactin and tumor growth factors) initiate their actions through such membrane receptors. Furthermore, since the growth and differentiation of the epithelial cells are often influenced by the behavior of the surrounding non-epithelial cells, diet induced changes in these cells may also indirectly affect the tumor development as well.[59,60] Mammary fibroblasts, for example, appear necessary for epithelial cells *in vitro* to maintain their responsiveness to estrogen with regard to enhancement of progesterone receptors, and mammary adipocytes may provide important paracrines necessary for epithelial cell growth. These adipocytes have a prolactin sensitive lipase which may influence mammary epithelial cell development by regulating the fatty acids available to them.

Experiments examining the effects of differing dietary levels of n-6 polyunsaturated fat on the lipid composition of the mammary tumor cell membrane have indicated that certain dietary changes do affect the saturation index of the membrane lipid. When the dietary intake of polyunsaturated fatty acids is greatly reduced, there is an increase in the saturation index of the membrane lipid which is associated with a reduction in both the prolactin binding capacity and growth of the tumor cells.[23,61] Subsequent increments in the dietary level of n-6 polyunsaturated lipid will initially increase the prolactin binding capacity and tumor growth, but once the dietary level of polyunsaturated fat has reached a certain threshold (e.g. 3% corn oil), no further increase in prolactin binding capacity occurs. This relationship is observed despite the fact that tumor development continues to correlate with the total amount of fat in the diet. Thus, it appears that once a specific minimum level of membrane desaturation has been reached which allows optimal receptor function, other lipid dependent mechanisms become more important to the regulation of tumor growth. This may explain the previously described requirement for a "basal" amount of polyunsaturated lipid before mammary tumor growth begins to correlate well with the total amount of fat in the diet.[18,22,62]

Another way dietary induced cellular lipid alterations may regulate tumor development is through their influence on the pool of precursor molecules available for eicosanoid metabolism. Malignant cells commonly synthesize more eicosanoids than do their benign counterparts,[63] and this increase may occur in both the cyclooxygenase and lipoxygenase pathways. The eicosanoid products generated by these tumor cells may influence tumor growth through the release of lysosomal enzymes from tumor cells to initiate invasion, the regulation of chemotactic migration of tumor cells in response to stimuli, and by affecting the adherence of tumor cells to biologically relevant substrata.[64] Nearly all tumor promoters stimulate arachidonic acid metabolism,[65] and in some circumstances, the responsible tumor growth factor is hormone dependent.[66] Finally, eicosanoids are also known to modulate immune function, which may have some effect on tumor immunosurveillance.[67]

The evidence that eicosanoid metabolites are involved in the promotion of mammary tumor growth by dietary lipids comes from several sources. One has been the data showing that pharmacological inhibitors of prostaglandin synthesis (e.g., indomethacin) can impair the enhancement of the mammary tumor development normally produced by increased levels of dietary n-6 lipid.[68,69] Perhaps more important from the dietary perspective, however, has been the observation that animals fed diets containing high dietary levels of omega-3 polyunsaturated fats, which selectively reduce the synthesis of certain eicosanoids, can inhibit mammary tumor development.[31,32,37] It has been convincingly demonstrated that both the normal and neoplastic membranes of such animals acquire increased proportions of eicosapentaenoic acid (C 20:5), a polyunsaturated fatty acid that is a known competitor of arachidonic acid (C 20:4) as a substrate for cyclooxygenases and lipoxygenases. This particular n-3 fatty acid has been reported to reduce the formation of the 2-series eicosanoids[70] and is also the precursor of the trienoic prostanoids and pentaene leukotrienes. Each of these eicosapentaenoic acid derived eicosanoids is less biologically active than the corresponding arachidonic acid derivative. It seems quite possible that such disturbances in eicosanoid metabolism could reduce tumor cell responsiveness to other promoters and improve cell-cell communication and allow better control of cell proliferation.[71]

Several laboratories were able to demonstrate convincingly that tumors from animals fed fish oil diets had reduced levels of tumor prostaglandin E, prostaglandin 6-keto-F_{1a}, and leukotriene B_4 relative to those from corn oil fed animals, but they were unable to specify which eicosanoid metabolites were directly responsible for the differences found in tumor development.[72-74] This issue has been addressed recently, however, in several interesting experiments by Abou-El-Ela *et al.*[75] Using DMBA treated rats they examined the eicosanoid metabolism in tumors derived from 6 diet/drug treatment groups: 20% corn oil; 20% corn oil + 0.004% indomethacin (an inhibitor of cyclooxygenase); 20% corn oil + 0.5% difluoromethylornithine (an inhibitor of ornithine decarboxylase); 20% corn oil + 0.004% indomethacin + 0.5% difluoromethylornithine; 15% menhaden oil + 5% corn oil; 15% menhaden oil + 5% corn oil + 0.5% difluoromethylornithine. The results indicated that the rats fed diets containing an elevated omega-3:omega-6 fatty acid ratio had reduced tumorigenesis, their tumors had diminished prostaglandin E and leukotriene B_4 production, and reduced ornithine decarboxylase activity. The inability of cyclooxygenase and ornithine decarboxylase inhibition alone, in the absence of lipoxygenase inhibition, to prevent mammary tumor promotion by 20% corn oil suggested that it was the alterations in leukotriene B_4 metabolism that were particularly important to tumor promotion in this model. It was, therefore, postulated that the inhibitory effects of high omega-3 polyunsaturated fatty acid diets on mammary tumorigenesis were more closely related to their antagonism of lipoxygenase activity than their antagonism of cyclooxygenase activity. This conclusion is also supported by recent studies on the *in vitro* growth characteristics of the MDA-MB-231 human breast cancer cell line. These leukotriene dependent cells are stimulated by linoleate, and inhibited by the omega-3 polyunsaturated fatty acids docosahexaenoic acid and eicosapentaenoic acid in a dose related fashion.[76] Similarly, Buckman et al.[77] have found that lipoxygenase products rather than cyclooxygenase metabolites play a major role in the linoleate-stimulated growth of mouse metastatic mammary tumor 4526 cells *in vitro*.

Whether these changes in leukotriene metabolism are sufficient to account for all of the differences in tumor development found between animals fed different polyunsaturated fat diets is yet to be determined. Recently, Gonzalez et al.[78] have reported on two human breast cancer cell lines which when transplanted into female nude mice fed high corn oil and fish oil diets, grew less well in the fish oil diet treated animals. Furthermore, this reduction in tumor development in one of the cell lines was accompanied by increased tumor lipid peroxidation levels. They also noted that supplementation of the fish oil diets with antioxidants reduced the tumor levels of the peroxidation products and significantly increased tumor volumes, while supplementation of the diet with ferric citrate, a peroxidant cation, significantly increased tumor lipid peroxidation product levels and decreased tumor volume. On the basis of these findings, they have proposed that the tumor growth suppressing activities of fish oil diets are due, at least in part, to the lipid peroxidation products. Finally, DeWille et al.[79] using MMTV/v-Ha-*ras* transgenic mice have shown that animals fed a 25% corn oil diet not only have an enhanced incidence of mammary tumors, but increased levels of *ras* mRNA. The specific reason for this enhancement is not yet defined, but it raises yet another possible mechanism by which lipids may influence tumor development.

IV. Concluding Remarks

When viewed collectively, this assortment of data from a variety of animal models strongly reenforces the hypothesis that both qualitative and quantitative changes in dietary lipid composition can affect breast cancer development. In almost all of these tumor models the data suggest that diets or culture media containing high levels of omega-6 polyunsaturated fatty acid stimulate mammary tumor development. Moreover, in some models this effect can still be expressed even when the diet alterations are begun several months after tumor initiation. There is convincing evidence that differences in the lipid composition of the diet are reflected in fatty acid alterations of both the neoplastic and non-neoplastic cell membranes which can importantly affect the cell's subsequent physiological and pathological behavior. The data suggest that these membrane alterations can significantly affect the cell's responsiveness to external stimuli (e.g. growth factors and hormones), its membrane permeability, and/or its selective enhancement of specific enzyme pathways which influence the cellular metabolism of certain carcinogens or chemotherapeutic agents. Moreover, since both membrane and intracellular fatty acids are substrates for a variety of intracellular metabolic events such as eicosanoid synthesis, changes in the fatty acid composition of these lipid pools may both directly influence neoplastic cell growth and differentiation and indirectly affect it through the induced effects on surrounding tissues and the immune system.

Currently, no unifying biochemical mechanism suffices to explain neatly all of the effects observed. Such a mechanism must adequately account for: 1) the selective nature of the dietary lipid effect on mammary cancer; 2) the requirement for a threshold amount of polyunsaturated lipid; 3) the contrasting differences between the n-3 and the n-6 polyunsaturated lipid families; and 4) the fact that the appropriate hormonal environment is required for its expression in hormone dependent tumors. Possibly, no single biochemical mechanism will be able to account for all the experimental variations in mammary tumor growth caused by qualitative and quantitative differences in dietary fat, just as no single hormone has been identified as the only endocrine factor responsible for the neoplastic transformation of the mammary gland. Ultimately, the response observed in any given dietary situation may reflect the net effect of several biochemical mechanisms. Certainly, a better understanding of these interrelationships is essential, however, if we are to effectively and appropriately apply dietary interventions in the prevention of human disease.

References

1. Tannenbaum, A., The genesis and growth of tumors. III. Effects of a high fat diet, *Cancer Res.* 2:468 (1942).
2. Tannenbaum, A., The dependence of tumor formation on the composition of the calorie restricted diet as well as the degree of restriction, *Cancer Res.* 5:616 (1945).
3. Tannenbaum, A., Nutrition and cancer, in: *The Physiopathology of Cancer*, F. Homburger, ed., Hoeber-Harper, New York (1959).
4. Silverstone, H. and A. Tannenbaum, The effect of the proportion of dietary fat on the rate of formation of mammary carcinoma in mice, *Cancer Res.* 10:448 (1950).
5. Davis, R.K., G.T. Stevenson, and K.A. Bausch, Tumor incidence in normal Sprague-Dawley female rats, *Cancer Res.* 16:194 (1956).
6. Carroll, K.K. and H.T. Khor, Effects of level and type of dietary fat on incidence of mammary tumors induced in female Sprague-Dawley rats by 7,12-dimethylbenz[a]anthracene. *Lipids* 6:415 (1971).
7. Welsch, C.W., Relationship between dietary fat and experimental mammary tumorigenesis: a review and a critique, *Cancer Res.* 52 (Suppl):2040s (1992).
8. Carroll, K.K., Experimental evidence of dietary factors and hormone-dependent cancers, *Cancer Res.* 35:3374 (1975).
9. Freedman, L.S., C. Clifford, and M. Messina, Analysis of dietary fat, calories, body weight and the development of mammary tumors in rats and mice: a review, *Cancer Res.* 50:5710 (1990).
10. Engel, R.W. and D.H. Copeland, Influence of diet on the relative incidence of eye, mammary, ear-duct, and liver tumors in rats fed 2-acetylaminofluorene, *Cancer Res.* 11:180 (1951).
11. Erickson, K.L., Dietary fat influences murine melanoma growth and lymphocyte-mediated cytotoxicity, *J. Natl. Cancer Inst.* 72:115 (1984).
12. Lavik, P.S. and C.A. Baumann, Further studies on the tumor promoting action of fat, *Cancer Res.* 3:749 (1943).
13. Tannenbaum, A., The dependence of tumor formation on the composition of the calorie-restricted diet as well as the degree of restriction, *Cancer Res.* 5:616 (1945).
14. Nigro, N.D., D.V. Singh, R.L. Campbell, and M.S. Pak, Effect of dietary beef fat on intestinal tumor formation by azoxymethane in rats, *J. Natl. Cancer Inst.* 54:439 (1975).
15. Roebuck, B.D., J.D. Yager Jr., and D.S. Longnecker, Dietary modulation of azaserine-induced pancreatic carcinogenesis, *Cancer Res.* 41:888 (1981).

16. Baumann, C.A., H.P. Jacobi, and H.P. Rusch, The effect of diet on experimental tumor production, *Am. J. Hyg.* 30A:1 (1939).

17. Lawrason, F.D. and A. Kirschbaum, Dietary fat with reference to spontaneous appearance and induction of leukemia in mice, *Proc. Soc. Exp. Biol. Med.* 56:6 (1944).

18. Carroll, K.K. and G.J. Hopkins, Dietary polyunsaturated fat versus saturated fat in relation to mammary carcinogenesis, *Lipids* 14:155 (1979).

19. Kidwell, W.R., M.E. Monaco, M.S. Wicha, and G.S. Smith, Unsaturated fatty acid requirements for growth and survival of a rat mammary tumor cell line, *Cancer Res.* 38:4091 (1978).

20. Wicha, M.S., L.A. Liotta, and W.R. Kidwell, Effects of free fatty acids on the growth of normal and neoplastic rat mammary epithelial cells, *Cancer Res.* 39:426 (1979).

21. Kidwell, W.R., R.A. Knazek, B.K. Vonderhaar, and T. Losonczy, *in: Molecular Interrelations of Nutrition and Cancer*, M.S. Arnott, J. van Eys, and Y.M. Wang, eds., Raven Press, New York (1982).

22. Hopkins, G.J. and K.K. Carroll, Relationship between the amount and type of dietary fat in promotion of mammary carcinogenesis induced by 7,12-dimethylbenzanthracene, *J. Natl. Cancer Inst.* 62:1009 (1979).

23. Cave, W.T., Jr. and J.J. Jurkowski, Dietary lipid effects on the growth, membrane composition, and prolactin-binding capacity of rat mammary tumors, *J. Natl. Cancer Inst.* 73:185 (1984).

24. Ip, C., C.A. Carter, and M.M. Ip, Requirement of essential fatty acid for mammary tumorigenesis in the rat, *Cancer Res.* 45:1997 (1985).

25. Hillyard, L., G.A. Rao, and S. Abraham, Effect of dietary fat and fatty acid on composition of mouse and rat mammary adenocarcinomas, *Proc. Soc. Exp. Biol. Med.* 163:376 (1980).

26. Carroll, K.K. and H.T. Khor, Effects of level and type of dietary fat on incidence of mammary tumors induced in female Sprague-Dawley rats by 7,12-dimethylbenzanthracene, *Lipids* 6:415 (1971).

27. Cohen, L.A., D.O. Thompson, Y. Maeura, K. Choi, M.E. Blank, and D.P. Rose, Dietary fat and mammary cancer. 1. Promoting effects of different dietary fats on *N*-nitrosomethylurea-induced rat mammary tumorigenesis, *J. Natl. Cancer Inst.* 77:33 (1986).

28. Cohen, L.A., D.O. Thompson, Y. Maeura, and J.H. Weisburger, Influence of medium chain triglycerides on the development of *N*-methylnitrosourea-induced rat mammary tumors, *Cancer Res.* 44:5023 (1984).

29. Lasekan, J.B., M.K. Clayton, A. Gendron-Fitzpatrick, and D.M. Ney, Dietary olive and safflower oils in promotion of DMBA-induced mammary tumorigenesis in rats, *Nutr. Cancer* 13:153 (1990).

30. Buckman, D.K., R.S. Chapkin, and K.L. Erickson, Modulation of mouse mammary tumor growth and linoleate enhanced metastasis by oleate, *J. Nutr.* 120:148 (1990).

31. Jurkowski, J.J. and W.T. Cave Jr., Dietary effects of menhaden oil on the growth and membrane lipid composition of rat mammary tumors, *J. Natl. Cancer Inst.* 74:1145 (1985).

32. Braden, L.M. and K.K. Carroll, Dietary polyunsaturated fat in relation to mammary carcinogenesis in rats, *Lipids* 21:285 (1986).

33. Tinsley, I.J., J.A. Schmitz, and D.A. Pierce, Influence of dietary fatty acids on the incidence of mammary tumors in the C3H mouse, *Cancer Res.* 41:1460 (1981).

34. Cave, W.T., Jr. and J.J. Jurkowski, Comparative effects of omega-3 and omega-6 dietary lipids on rat mammary tumor development, *in: Proceedings of the AOCS Short Course on Polyunsaturated Fatty Acids and Eicosanoids*, W.F.M. Lands, ed., American Oil Chemists' Society, Champaign (1987).

35. Ip, C., M.M. Ip, and P. Sylvester, Relevance of *trans* fatty acids and fish oil in all tumorigenesis studies, *in: Dietary Fat and Cancer*, C. Ip, D.F. Birt, A.E. Rogers, and C. Mettlin, eds., Alan R. Liss, New York (1986).

36. Karmali, R.A., R.U. Doshi, L. Adams, and K. Choi, Effect of n-3 fatty acids on mammary tumorigenesis, *Adv. Prostaglandin Thromboxane Leukot. Res.* 17:886 (1987).

37. Karmali, R.A., J. Marsh, and C. Fuchs, Effect of omega-3 fatty acids on growth of a rat mammary tumor, *J. Natl. Cancer Inst.* 73:457 (1984).

38. Gabor, H. and S. Abraham, Effect of dietary menhaden oil on tumor cell loss and the accumulation of mass of a transplantable adenocarcinoma in BALB/c mice, *J. Natl. Cancer Inst.* 76:1223 (1986).

39. Kort, W.J., I.M. Weijma, A.J. Vergroesen, and D.L. Westbroek, Conversion of diets at tumor induction shows the pattern of tumor growth and metastasis of the first given diet, *Carcinogenesis* 8:611 (1987).

40. Pritchard, G.A., D.L. Jones, and R.E. Mansel, Lipids in breast carcinogenesis, *Br. J. Surg.* 76:1069 (1989).

41. Blank, E.W. and R.L. Ceriani, Fish oil enhancement of ^{131}I-conjugated anti-human milk fat globule monoclonal antibody experimental radioimmunotherapy of breast cancer, *J. Steroid Biochem.* 34:149 (1989).

42. Gabor, H., E.W. Blank, and R.L. Ceriani, Effect of dietary fat and monoclonal antibody therapy on the growth of human mammary adenocarcinoma MX-1 grafted in athymic mice, *Cancer Lett.* 52:173 (1990).

43. Borgeson, C.E., L. Pardini, R.S. Pardini, and R.C. Reitz, Effects of dietary fish oil on human mammary carcinoma and on lipid-metabolizing enzymes, *Lipids* 24:290 (1989).

44. Berenblum, I., Challenging problems in carcinogenesis, *Cancer Res.* 45:1917 (1985).

45. Barbacid, M., Oncogenes and human cancer: cause or consequence?, *Carcinogenesis* 7:1037 (1986).

46. Carroll, K.K. and H.T. Khor, Effects of dietary fat and dose level of 7,12-dimethylbenzanthracene on mammary tumor incidence in rats, *Cancer Res.* 30:2260 (1970).

47. Hopkins, G.J. and C.E. West, Effect of dietary polyunsaturated fat on the growth of transplantable C3HA'yfB mice, *J. Natl. Cancer Inst.* 58:753 (1977).

48. Abraham, S. and G.A. Rao, *in: Control Mechanisms in Cancer*, W.E. Criss, T. Ono, and J.R. Sabine, eds., Raven Press, New York (1976).

49. Rao, G.A. and S. Abraham, Brief communication: enhanced growth rate of transplanted mammary adenocarcinoma induced in C3H mice by dietary linoleate, *J. Natl. Cancer Inst.* 56:431 (1976).

50. Ip, C. and D. Sinha, Neoplastic growth of carcinogen-treated mammary transplants as influenced by fat intake of donor and host, *Cancer Lett.* 11:277 (1981).

51. Ip, C., Ability of dietary fat to overcome the resistance of mature female rats to 7,12-dimethylbenzanthracene-induced mammary tumorigenesis, *Cancer Res.* 40:2785 (1980).

52. Aylsworth, C.F., C. Jone, J.E. Trosko, J. Meites, and C.W. Welsch, Promotion of 7,12-dimethylbenzanthracene-induced mammary tumorigenesis by high dietary fat in the rat: possible role of intercellular communication, *J. Natl. Cancer Inst.* 72:637 (1984).

53. Welsch, C.W. and H. Nagasawa, Prolactin and murine mammary tumorigenesis: a review, *Cancer Res.* 37:951 (1977).

54. Dao, T.L. and R. Hilf, Dietary fat and breast cancer: a search for mechanisms, *Adv. Exp. Med. Biol.* 322:223 (1992).

55. Chan, P.C. and L.A. Cohen, Effect of dietary fat, antiestrogen, and antiprolactin on the development of mammary tumors in rats, *J. Natl. Cancer Inst.* 52:25 (1974).

56. Aylsworth, C.F., D.A. Van Vugt, P.W. Sylvester, and J. Meites, Role of estrogen in the promotion of dimethylbenzanthracene-induced mammary tumors by dietary fat, *Cancer Res.* 44:2835 (1984).

57. Wetsel, W.C., A.E. Rogers, A. Rutledge, and W.W. Leavitt, Absence of an effect of dietary corn oil content on plasma prolactin, progesterone, and 17β-estradiol in female Sprague-Dawley rats, *Cancer Res.* 44:1420 (1984).

58. Ip, C., P. Yip, and L.L. Bernardis, Role of prolactin in the promotion of dimethylbenzanthracene-induced mammary tumors by dietary fat, *Cancer Res.* 40:374 (1980).

59. Oka, T. and M. Yoshimura, Paracrine regulation of mammary gland growth, *Clin. Endocrinol. Metab.* 15:79 (1986).

60. Carroll, K.K. and H.I. Parenteau, A proposed mechanism for effects of diet on mammary cancer, *Nutr. Cancer* 16:79 (1991).

61. Cave, W.T., Jr. and M.J. Erickson-Lucas, Effects of dietary lipids on lactogenic hormone receptor binding in rat mammary tumors, *J. Natl. Cancer Inst.* 68:319 (1982).

62. Hopkins, G.J., T.G. Kennedy, and K.K. Carroll, Polyunsaturated fatty acids as promoters of mammary carcinogenesis induced in Sprague-Dawley rats, *J. Natl. Cancer Inst.* 66:517 (1981).

63. Jaffe, B.M. and M.G. Santaro, Prostaglandin production by tumors, *in*: *Prostaglandins in Cancer Research*, E. Geraci, P. Rudolfo, and S. Santaro, eds., Springer-Verlag, New York (1987).

64. Dunn, J.R. II, D.W. Ohannesian, S. Tefend, D. Malick, A. Kendall, J.D. Taylor, B.F. Sloane, and K.V. Honn, The effects of eicosanoid inhibitors on tumor cell arachidonic acid (and metabolite) uptake, release, and metabolism, *in*: *Prostaglandins in Cancer Research*, E. Geraci, P. Rudolfo, and S. Santaro, eds., Springer-Verlag, New York (1987).

65. Levine, L., Tumor promoters, growth factors, and arachidonic acid metabolism, *in*: *Prostaglandins in Cancer Research*, E. Geraci, P. Rudolfo, and S. Santaro, eds., Springer-Verlag, New York (1987).

66. Ravidin, P.M. and V.C. Jordan, Peptide growth factors, estrogens, and anti-estrogens: integrated effects on the proliferation and differentiation of normal and neoplastic breast tissue, *in*: *Growth Factors and Oncogenes in Breast Cancer*, M. Sluyser, ed., Ellis Horwood, Chichester (1987).

67. Favalli, C., A. Mastino, and E. Geraci, Prostaglandins in immunotherapy of cancer, *in*: *Prostaglandins in Cancer Research*, E. Geraci, P. Rudolfo, and S. Santaro, eds., Springer-Verlag, New York (1987).

68. Hillyard, L.A. and S. Abraham, Effect of dietary polyunsaturated fatty acids on the growth of mammary adenocarcinomas in mice and rats, *Cancer Res.* 39:4430 (1979).

69. Carter, C.A., R.J. Millholland, W. Shea, and M.M. Ip, Effect of the prostaglandin synthase inhibitor indomethacin on 7,12-dimethylbenzanthracene-induced mammary tumorigenesis in rats fed different levels of fat, *Cancer Res.* 43:3559 (1983).

70. Culp, B.R., B.G. Titus, and W.E. Lands, Inhibition of prostaglandin biosynthesis by eicosapentaenoic acid, *Prostaglandins Med.* 3:269 (1979).

71. Wade, A.E. and O.R. Bunce, Influences of dietary omega-3 fatty acids on drug and carcinogen metabolism and on the promotion of DMBA-induced mammary tumors, *in*: *Polyunsaturated Fatty Acids and Eicosanoids*, W.E.M. Lands, ed., American Oil Chemists' Society, Champaign (1987).

72. Abou-El-Ela, S., K.W. Prasse, R. Carroll, A.E. Wade, S. Dharwadkar, and O.R. Bunce, Eicosanoid synthesis in 7,12-dimethylbenz[a]anthracene-induced mammary carcinomas in Sprague-Dawley rats fed primrose oil, menhaden oil, or corn oil diet, *Lipids* 23:948 (1988).

73. Karmali, R.A., A. Donner, S. Gobel, and T. Shimamura, Effect of n-3 and n-6 fatty acids on 7,12 dimethylbenz[a]anthracene-induced mammary tumorigenesis, *Anticancer Res.* 9:1161 (1989).

74. Takata, T., T. Minoura, H. Takada, M. Sakaguchi, M. Yamamura, K. Hioki, and M. Yamamoto, Specific inhibitory effect of dietary eicosapentaenoic acid on *N*-nitroso-*N*-methyl-urea-induced mammary carcinogenesis in female Sprague-Dawley rats, *Carcinogenesis* 11:2015 (1990).

75. Abou-El-Ela, S., K.W. Prasse, R. Carroll, A.E. Wade, and O.R. Bunce, Effects of D,L-2-difluoromethylornithine and indomethacin on mammary tumor promotion in rats fed high n-3 and/or n-6 fat diets, *Cancer Res.* 49:1434 (1989).

76. Rose, D.P. and J.M. Connolly, Effects of fatty acids and inhibitors of eicosanoid synthesis on the growth of a human breast cancer cell line in culture, *Cancer Res.* 50:7139 (1990).

77. Buckman, D.K., N.E. Hubbard, and K.L. Erickson, Eicosanoids and linoleate-enhanced growth of mouse mammary tumor cells, *Prostaglandins Leukot. Essent. Fatty Acids* 44:177 (1991).

78. Gonzalez, M.J., R.A. Schemmel, J.I. Gray, L. Dugan Jr., L.G. Sheffield, and C.W. Welsch, Effect of dietary fat on growth of MCF-7 and MDA-MB231 human breast carcinomas in athymic nude mice: relationships between carcinoma growth and lipid peroxidation product levels, *Carcinogenesis* 12:1231 (1991).

79. DeWille, J.W., K. Waddell, C. Steinmeyer, and S.J. Farmer, Dietary fat promotes mammary tumorigenesis in MMTV/v-Ha-*ras* transgenic mice, *Cancer Lett.* 69:59 (1993).

Chapter 6
Effect of Conjugated Linoleic Acid on Carcinogenesis

JOSEPH A. SCIMECA, HENRY J. THOMPSON, and CLEMENT IP

I. Introduction

Conjugated linoleic acid (CLA) refers to a mixture of positional and geometric isomers of linoleic acid. The two double bonds in CLA are primarily in positions 9 and 11, or 10 and 12, along the carbon chain, thus giving rise to the designation of a conjugated diene. Each of the double bonds can be in the *cis* or *trans* configuration, and hence eight isomers are formed from the isomerization of linoleic acid.

CLA is a naturally occurring substance and was initially identified as an antimutagenic agent from grilled ground beef.[1,2] Subsequent investigations revealed CLA to be present in a variety of foods.[3-5] In general, CLA was higher in meats derived from ruminants than non-ruminants (values ranged from 0.6 to 5.6 mg CLA/g fat). Seafood contained little CLA. In contrast, dairy products contained considerable, but variable amounts of CLA. For example, the CLA content in natural and process cheese ranged from 3 to 9 mg CLA/g fat. The fact that considerable differences occur in the CLA content of common foods suggests that dietary intake of CLA may vary widely, depending on an individual's food selections.

Despite evidence from Pariza's laboratory in 1987 that CLA is antimutagenic, a surprisingly meager amount of in-depth investigation has been conducted since that time. This paper will nevertheless provide a brief review of those studies which relate the biological effects of CLA to various aspects of carcinogenesis.

II. Conjugated Linoleic Acid and Carcinogenesis

It is not the intent of this paper to provide a general review of the scientific evidence linking certain biological events to carcinogenesis, nor is it intended to provide a unified theory explaining the mechanism of action of CLA in carcinogenesis. Rather, this review will simply use well-accepted biomarkers associated with carcinogenesis to discuss the

---

Joseph A. Scimeca ● Nutrition Department, Kraft General Foods, Glenview, Illinois; Henry J. Thompson ● Laboratory of Nutrition Research, AMC Cancer Research Center, Denver, Colorado; Clement Ip ● Department of Surgical Oncology, Roswell Park Cancer Institute, Buffalo, New York

Diet and Breast Cancer, Edited under the auspices of the
American Institute for Cancer Research, Plenum Press, New York, 1994

anticarcinogenic potential of CLA. This evidence is summarized in Table 1, and is discussed in more detail in the following text.

A. Tumorigenesis

Pariza, Ha, and coworkers were the first investigators to conclusively demonstrate that CLA had anticarcinogenic activity. Using the two-stage mouse epidermal carcinogenesis model, synthetically prepared CLA was topically applied at 7 days (20 mg/mouse), 3 days (20 mg/mouse), and 5 minutes (10 mg/mouse) prior to treatment with 50 nmol/mouse of 7,12-dimethylbenz[a]anthracene (DMBA).[2] Commencing one week later, the female mice (CD-1) received twice weekly topical applications of 12-*O*-tetradecanoylphorbol-13-acetate (TPA) until study termination. Results indicated that CLA-treated mice had a lower incidence (~ 15% reduction) and approximately 50% fewer papillomas/mouse than did controls (which were painted with linoleic acid).

In a follow-up study by the same investigators, synthetic CLA inhibited benzo[a]pyrene-induced forestomach tumors in female ICR mice.[6] CLA (0.1 ml) was administered by gavage 4 and 2 days prior to each weekly oral administration of 2 mg benzo[a]pyrene (BP), for a 4 week period. In three separate experiments, CLA reduced the number of neoplasms by about 50% over controls (0.1 ml linoleic acid), as well as reducing tumor incidence (10-21%).

Table 1. CLA Effect on Events Associated with Carcinogenesis

Event	Species/ Cell Type	Experimental Model	Rte. of Admin.	Major Finding	Ref.
Tumorigenesis	Mouse	DMBA skin tumors.	Dermal	Inhibition in incidence and multiplicty.	2
	Mouse	BP forestomach tumors.	Gavage	Inhibition in incidence and multiplicty.	6
	Rat	DMBA breast tumors.	Diet	Dose-response inhibition in incidence, multiplicity and burden.	7
	Rat	AOM intestinal tumors.	Diet	No effect.	8
	Rat	DMBA and MNU breast tumors.	Diet	Inhibition w/short-term feeding.	10
Promotion biomarker	Mouse	Forestomach ODC.	Gavage	Inhibition of TPA-induced ODC activity.	11
Mitogenesis	Rat	BrdU labeling of mamary gland.	Diet	Inhibition of lobulo-alveolar proliferation.	10
	M21-HPB, HT-29, MCF-7	Cell growth curves.	*In vitro*	Cytostatic and cytotoxic effects at micromolar concentrations.	12, 13
Mutagenesis	Salmonella	Ames assay.	Gavage	CLA incorporated into S-9 inhibits IQ activation, but not BP or DMBA activation.	14
Carcinogen activation & detoxification	Mouse	IQ-DNA adducts.	Gavage	Inhibition in certain target (liver & lung) and non-target organs, (lg intestine & kidney), but inactive in other target organs (stomach) and non-target organs (sm intestine).	15
	Rat	Mamary and liver phase II enzymes.	Diet	No effect on liver and mammary GST, or liver UDP-GT activity.	7
Signal transduction	Mouse	Forestomach extracts.	Gavage	PKC-like activity refractory to activation.	17
	3T3	Phospholipid turnover.	*In vitro*	Modulation of TPA-stimulated phospholipase C activity.	18
Antioxidant capacity	Rat	Ex vivo TBARS.	Diet	Inhibition in mammary gland, but not liver.	7

Abbreviations: DMBA, 7, 12-dimethylbenz[a]anthracene; BP, benzo[a]pyrene; AOM, azoxymethane; MNU, methylnitrosourea; IQ, 2-amino-3-methylimidazo[4,5-f] quinoline; ODC, ornithine decarboxylase; TPA, 12-*O*-tetradecanoylphorbol-13-acetate; BrdU, bromodeoxyuridine; PKC, protein kinase C; GST, glutathione-S-transferase; UDP-GT, UDP-glucuronyl transferase; TBARS, thiobarbituric acid reactive substances.

In contrast to the above mentioned acute dosing regimen in which experimental animals were intubated with CLA, Ip and associates fed rats an AIN-76A basal diet, or the same diet supplemented with 0.5%, 1.0%, or 1.5% synthetically prepared CLA.[7] Female Sprague-Dawley rats were placed on the diets two weeks before treatment with DMBA (10 mg i.g.), and continued on their diets until study termination 24 weeks after the carcinogen administration. Results revealed a dose-dependent reduction in the total number of mammary tumors, with the magnitude of inhibition ranging from 32-60%. Tumor multiplicity was similarly reduced by 33-60%, and tumor incidence by 17-50%.

A parallel study, using similar experimental diets (including the same source of synthetic CLA), was conducted to examine the effect of dietary CLA on intestinal tumorigenesis.[8] Male Fischer 344 rats were fed an AIN-76A basal diet, or the same diet supplemented with 0.5%, 1.0%, or 1.5% CLA, for 36 weeks. Two weeks after the start of the study, the rats received weekly subcutaneous injections with azoxymethane (AOM) for the following three weeks (15 mg/kg body weight per week). Results indicated that the incidence and the multiplicity of intestinal tumors were not affected by CLA (Table 2). Similarly, tumor size and location within the intestinal tract were not influenced by the CLA (data not shown). This finding that CLA was ineffective as an anticarcinogen in this model is not surprising given the variable results seen by other investigators examining the effect of dietary fat on animal colon carcinogenesis. Newberne and Nauss determined that of the 28 studies of experimental colon carcinogenesis they evaluated, 12 studies found no effect of high levels of fat.[9] The authors of this review concluded that dietary variables other than fat level have an effect on an animal's response to a colon carcinogen. This may help explain the inconsistencies in the literature regarding the effect of dietary fat in the carcinogen-induced model of colon carcinogenesis.

In order to characterize further the effect of CLA on rodent mammary cancer, we have recently completed additional investigations using lower levels of CLA.[10] Using the same DMBA model, rats were fed diets containing CLA at concentrations of 0.05%, 0.1%, 0.25%, and 0.5%, starting two weeks before DMBA (5 mg i.g.) and continuing for 9 months. This study lasted longer than the initial one because the tumors took a longer time to develop with the low dose of DMBA. However, even at these lower levels of CLA, there was a dose-dependent inhibition of mammary tumors, with total mammary tumor yield being reduced from 22-58%. Inter-group comparison revealed that 0.1% dietary CLA was sufficient to produce a significant inhibition of mammary tumors.

In a follow-up experiment, we examined the effect of short-term CLA feeding on rat mammary tumorigenesis. Female rats were fed 1.0% CLA from weaning to about 50 days of age, a time which corresponds to maturation of the mammary gland.[10] Two different carcinogens were administered to induce mammary tumors: DMBA (10 mg, i.g.), which

Table 2. Incidence and Multiplicity of AOM-Induced Intestinal Tumors

		No. of Animals with Intestinal Tumors			Mean No. of Tumors Per Rat		
Diet	No. of Rats	Small Intestine	Large Intestine	Total	Small Intestine	Large Intestine	Total
Control	33	27 (82)[a]	15 (45)	31 (94)	1.15	0.76	1.91
0.5% CLA	32	24 (75)	13 (41)	30 (94)	1.13	0.59	1.72
1.0% CLA	33	27 (82)	16 (48)	32 (97)	1.30	0.73	2.03
1.5% CLA	35	27 (77)	16 (46)	30 (86)	1.17	0.49	1.66

[a] Numbers in parenthesis, percentage of animals with tumors.

requires metabolic activation, and methylnitrosourea (MNU) (6 mg i.p.), which is a direct acting alkylating agent. Results showed that dietary supplementation for 5 weeks with CLA significantly inhibited total mammary tumor yield by 39% and 34% in the DMBA and MNU models, respectively. Importantly, the fact that CLA inhibited MNU-induced tumors suggests that it may modulate the susceptibility of the target organ to carcinogenesis. This study also suggests that the timing of the CLA exposure may be important.

B. Promotion Biomarker

The possibility that CLA may influence the promotional phase of carcinogenesis was revealed in a study conducted by Pariza and coworkers.[11] They examined the effect of CLA on mouse forestomach TPA-induced ornithine decarboxylase (ODC) activity. ODC activity is well known to be characteristically increased following treatment with tumor promoters. Female ICR mice were gavaged with CLA (100 mg) twice a week, for 1, 2, or 4 weeks prior to the TPA exposure. Results showed that following 4 weeks of CLA treatment, the peak ODC activity (6 hrs after exposure to TPA) was reduced 80% from control (TPA but no CLA). Hence, this study provides supporting evidence of the ability of CLA to inhibit tumor promotion.

C. Mitogenesis

Increased mitogenesis is a prerequisite for the expansion of transformed cells into malignancy. Most of the research in this area involving CLA has centered on the *in vitro* effect of CLA on growth and proliferation of cancer cell lines. The sole exception has been a preliminary study in which we examined the effect of dietary CLA on the *in vivo* incorporation of bromodeoxyuridine (BrdU) by rat mammary epithelial cells.[10] BrdU incorporation into normal proliferating cells was quantitated by immunostaining. Short-term feeding of 1% CLA diet from weaning to 55 days of age produced a 25% reduction in the proliferative activity of the lobulo-alveolar compartment of the mammary tree.

Other studies have shown that CLA inhibits the cell growth of a number of cancer cell lines.[12,13] CLA was incubated with three human cancer cell lines (M21-HPB, malignant melanoma; HT-29, colorectal; MCF-7 breast), at various physiological concentrations (1.8 to 7.1 micromolar). Results indicated that CLA produced a significant reduction in proliferation (18-100%) compared to controls. Cell viability was dose- and time-dependent. CLA was particularly effective against human breast cancer cells (MCF-7), producing a completely cytotoxic effect after 12 days of incubation at a concentration of 3.6 micromolar.

D. Mutagenesis

Pariza and coworkers initially determined that the antimutagenic activity present in crude extracts from grilled beef was due to CLA.[2] Subsequent investigations using the Ames assay found that CLA, when incorporated into mouse liver microsomal membranes (i.e., the S-9 fraction), selectively inhibits the activation of 2-amino-3-methylimidazo[4,5-*f*] quinoline (IQ).[14] Female ICR mice were gavaged with 0.1 ml olive oil (control), 0.1 ml olive oil plus 0.1 ml CLA, or 0.1 ml olive oil plus 0.1 ml linoleic acid, twice weekly for four weeks. At study termination, liver S-9 fractions were prepared. The S-9 from mice treated with CLA produced a significant reduction in the activation of IQ (at concentrations from 1-20 ng/plate). At the concentration of 20 ng IQ per plate, the activation was reduced by nearly 50%. In contrast, activation of BP or DMBA was unaffected. Pariza and coworkers concluded that CLA has a selective modulating effect on cytochrome P450 isozymes responsible for activating IQ, but not those which activate BP or DMBA.

E. Carcinogen Activation and Detoxification

Zu and Schut examined the effect of synthetic CLA on IQ-DNA adduct formation in CDF_1 mice.[15] IQ, a carcinogen derived from cooked meat, has the ability to produce tumors in mice in liver, forestomach, and lungs.[16] CLA was administered by gavage every other day for a period of 45 days (50 µl/48 hr for days 1-24, and 100 µl/48 hr for days 25-45). Control mice received trioctanoin. On day 46 all animals were gavaged with a single IQ dose (50 mg/kg), and tissues were collected 24 hr later. Tissue IQ-DNA adducts were determined by ^{32}P-postlabelling. Differential effects were noted for female and male mice. CLA pretreatment inhibited IQ-DNA adduct formation in both target organs (liver in males and females, lung in females) and non-target organs (large intestine and kidney in females). Particularly noteworthy was a nearly complete inhibition (>95%) of adduct formation in the female kidney. In contrast, CLA was found to be inactive in another target organ (stomach), and non-target organ (small intestine). Therefore, it appears that CLA may have the ability to reduce markedly IQ-DNA adduct formation leading to the initiation of tumors in certain organs.

The effect of CLA on phase II detoxification enzymes has only been investigated to a limited extent. Ip and associates fed female rats diets containing various levels of CLA for 1 month.[7] Data indicated that the CLA had no effect on glutathione-S-transferase activity in either the liver or mammary gland. These investigators also found no effect on liver UDP-glucuronyl transferase activity. Hence, it appears that dietary CLA has no effect on these particular phase II detoxifying enzymes, but may affect carcinogen metabolism via a modulation of selective phase I detoxifying enzymes (i.e., cytochrome P450 isozymes).

F. Signal Transduction

Effects of CLA on signal transduction as it relates to carcinogenesis have not been systematically studied. However, research from Pariza's laboratory has revealed some tantalizing findings. The protein kinase C (PKC) series of proteins are known to be involved in signal transduction and are thought to act as receptors for TPA. Pariza and coworkers have found that PKC-like activity, from partially purified extracts of forestomach tissue obtained from mice given CLA, was refractory to activation by TPA and phosphatidylserine in the absence of calcium.[17] Related to this was another study in which 3T3 cells were incubated with physiological concentrations of CLA or linoleic acid and the effect of CLA on phospholipase C activity was determined.[18] CLA was more effective than linoleic acid at modulating TPA-stimulated phospholipase C activity. These findings may eventually lead to an increased understanding of the mechanisms of the anticarcinogenic effects of CLA.

G. Antioxidant Capacity

The meager amount of evidence indicating an effect of CLA on oxidation is included due to the importance of oxidative events in carcinogenesis. Pariza and associates found CLA to be a potent *in vitro* antioxidant: at a molar ratio of 1 part CLA to 1000 parts linoleic acid, peroxide formation was inhibited by more than 90%.[6] Ip *et al.*[7] found that CLA had no effect on the amount of thiobarbituric acid reactive substances in the liver, but a significant decrease was found in the mammary gland. Interestingly, there was no dose-response relationship in the dietary range of 0.25 to 1.5% CLA, with all doses tested producing a 30-40% inhibition. The authors were unable to offer an explanation for the apparent insensitivity of the liver to CLA.

III. Discussion

CLA is a unique anticarcinogen because it is a naturally occurring substance primarily found in food products derived from animal sources. The fact that CLA is a fatty acid with potent anticarcinogenic action makes it even more novel. By way of comparison, fish oil is a lipid with a reported ability to inhibit tumorigenesis.[19,20] However, the amount of fish oil required to exert this effect usually exceeds 10% in the diet. As described in this paper, a dietary level of 0.1% CLA, or a hundred-fold less than that for fish oil, is sufficient to produce a significant inhibition in rat mammary tumors.

The designation of CLA as a potent animal anticarcinogen is supported by the experimental evidence. CLA was effective in inhibiting tumorigenesis in four separate models of chemically-induced carcinogenesis. Dietary CLA administration at a level of 0.1% for as short as five weeks produced significant inhibition of rat mammary tumor yield.

Additional evidence supporting the contention that CLA is an anticarcinogen comes from a broad array of studies. Effects of CLA on mitogenesis, mutagenesis, and carcinogen metabolism are especially interesting because they demonstrate a specificity of action. Generally these studies employ CLA at levels that are physiologically relevant. Particularly noteworthy is the study by Zu and Schut[15] with the food-derived carcinogen, IQ. Their work suggests other possible target sites (liver, lung and kidney) for further investigation into the anticarcinogenic action of CLA. Also relevant is the work by Ip and coworkers which has identified a critical "window" during the maturing of the rat mammary gland that appears to be particularly receptive to the antitumorigenic action of CLA. Feeding CLA at this time (from weaning to 55 days of age) produces a significant reduction in the proliferative activity in the lobulo-alveolar compartment of the mammary tree. It is conceivable that the antitumorigenicity of CLA could be due in part to an inhibition in the clonal expansion of the initiated mammary epithelial cells. Further investigation at the cellular and subcellular level should permit more definitive conclusions regarding the mechanisms of the anticarcinogenic action of CLA.

References

1. Pariza, M.W., L.J. Loretz, J.M. Storkson, and N.C. Holland, Mutagens and modulator of mutagenesis in fried ground beef, *Cancer Res.* (Suppl) 43:2444s (1983).
2. Ha, Y.L., N.K. Grimm, and M.W. Pariza, Anticarcinogens from fried ground beef: heat-altered derivatives of linoleic acid, *Carcinogenesis* 8:1881 (1987).
3. Ha, Y.L., N.K. Grimm, and M.W. Pariza, Newly recognized anticarcinogenic fatty acids: identification and quantification in natural and processed cheeses, *J. Agric. Food Chem.* 37:75 (1989).
4. Chin, S.F., W. Liu, J.M. Storkson, Y.L. Ha, and M.W. Pariza, Dietary sources of conjugated dienoic isomers of linoleic acid, a newly recognized class of anticarcinogens, *J. Food. Compos. Anal.* 5:185 (1992).
5. Shantha, N.C., E.A. Decker, and Z. Ustunol, Conjugated linoleic acid concentration in processed cheese, *J. Am. Oil Chem. Soc.* 69:425 (1992).
6. Ha, Y.L., J. Storkson, and M.W. Pariza, Inhibition of benzo[a]pyrene-induced mouse forestomach neoplasia by conjugated dienoic derivatives of linoleic acid, *Cancer Res.* 50:1097 (1990).
7. Ip, C., S.F. Chin, J.A. Scimeca, and M.W. Pariza, Mammary cancer prevention by conjugated linoleic acid, *Cancer Res.* 51:6118 (1991).
8. Scimeca, J.A., unpublished.
9. Newberne, P.M. and K.M. Nauss, Dietary fat and colon cancer: variable results in animal models, *in: Dietary Fat and Cancer*, C. Ip, D.F. Birt, A.E. Rogers, and C. Mettlin, eds., Alan R. Liss, New York (1986).

10. Ip, C., M. Singh, H.J. Thompson, and J.A. Scimeca, unpublished.
11. Benjamin, H., J.M. Storkson, K. Albright, and M.W. Pariza, TPA-mediated induction of ornithine decarboxylase activity in mouse forestomach and its inhibition by conjugated dienoic derivatives of linoleic acid, *FASEB J.* 4:A508 (1990).
12. Schultz, T.D., B.P. Chew, and W.R. Seaman, Differential stimulatory and inhibitory responses of human MCF-7 breast cancer cells to linoleic acid and conjugated linoleic acid in culture, *Anticancer Res.* 12:2143 (1992).
13. Schultz, T.D., B.P. Chew, W.R. Seaman, and L.O. Luedecke, Inhibitory effect of conjugated dienoic derivatives of linoleic acid and β-carotene on the *in vitro* growth of human cancer cells, *Cancer Lett.* 63:125 (1992).
14. Pariza, M.W., Y.L. Ha, J. Storkson, K. Albright, T. Lin, and H. Benjamin, Effect of CLA conjugated dienoic derivatives of linoleic acid on activation of IQ, 2-amino-3-methyl-imidazo-[4,5-*f*] quinoline for mutagenesis, *Proc. Am. Assoc. Cancer Res.* 32:128 (1991).
15. Zu, H.-X. and H.A.J. Schut, Inhibition of 2-amino-3-methylimidazo-[4,5-*f*] quinoline (IQ)-DNA adduct formation in CDF_1 mice by heat-altered derivatives of linoleic acid, *Food Chem. Toxicol.* 30:9 (1992).
16. Ohgaki, H., K. Kusama, N. Matsukura, K. Morino, H. Hasegawa, S. Sato, S. Takayama, and T. Sugimura, Carcinogenicity in mice of a mutagenic compound, 2-amino-3-methylimi-dazo[4,5-*f*]quinoline, from broiled sardine, cooked beef and beef extract, *Carcinogenesis* 5:921 (1984).
17. Benjamin, H., J.M. Storkson, W. Liu, and M.W. Pariza, The effect of conjugated dienoic derivatives of linoleic acid (CLA) on mouse forestomach protein kinase C (PKC)-like activity, *FASEB J.* 6:A1396 (1992).
18. Bonordon, W., J. Storkson, W. Liu, K. Albright, and M. Pariza, Fatty acid inhibition of 12-0-tetradecanoyl-phorbol 13-acetate (TPA)-induced phospholipase C activity, *FASEB J.* 7:A618 (1993).
19. Reddy, B.S., Omega-3 fatty acids as anticancer agents, in: *Health Effects of Dietary Fatty Acids*, G.J. Nelson, ed., American Oil Chemists' Society, Champaign (1991).
20. Cave, W.T. Jr., ω3 fatty acid diet effects on tumorigenesis in experimental animals, in: *Health Effects of ω3 Polyunsaturated Fatty Acids in Seafoods*, A.P. Simopoulous, R.R. Kifer, R.E. Martin, and S.M. Barlow, eds., Karger, Basel (1991).

A Possible Mechanism by Which Dietary Fat Can Alter Tumorigenesis: Lipid Modulation of Macrophage Function

KENT L. ERICKSON and NEIL E. HUBBARD

I. Introduction

Numerous experimental and epidemiological studies have provided evidence linking dietary fat with increased risk for breast cancer. Some epidemiological studies have reported a positive correlation between breast cancer and dietary fat intake[1] while a few have reported that no correlation existed.[2-5] In contrast, studies with animal models of mammary tumorigenesis are more consistent. In general, the studies in rodents showed that high levels of dietary fat led to an increased incidence of spontaneous or carcinogen-induced breast tumors as compared to animals fed a moderate or low level of dietary fat.[6] In addition, rodents fed diets containing polyunsaturated vegetable fats developed more tumors than animals fed saturated fats.[7,8] Not only has dietary fat been linked to altered primary tumor growth, but it also appears to influence the process of metastasis.[9] Since linoleic acid (18:2n-6) was the most abundant fatty acid found in many of the polyunsaturated vegetable oils, numerous investigators have suggested that it may be pivotal in the promotion of mammary tumorigenesis. The role of linoleic acid in increasing carcinogen-induced mammary tumor incidence[10] as well as metastasis[11] has been previously demonstrated.

Although the exact mechanisms by which dietary fat promotes mammary tumorigenesis are not known, there is experimental evidence to support a number of possibilities. Some of those possible mechanisms include: alteration in immunity, modulation of eicosanoid production, e.g. prostaglandins, leukotrienes, and hydroxy fatty acids, production of peroxides, changes in membrane fluidity or microviscosity, alteration in energy metabolism, alteration in hormones secreted, as well as several others (Table 1). We have focused on alteration in immune function because it can be concluded from a number of published reports that dietary fat can modulate certain, but not all components of an immune response. For example, lysis of mammary tumors by cytotoxic T cells was suppressed by high levels

---

Kent L. Erickson and Neil E. Hubbard ● Department of Cell Biology and Human Anatomy, University of California School of Medicine, Davis, California

Diet and Breast Cancer, Edited under the auspices of the American Institute for Cancer Research, Plenum Press, New York, 1994

Table 1. Possible Mechanisms by Which Dietary Fat Could Modify Tumor Growth

1. Alteration in immune responsiveness

2. Modulation of the type and quantity of eicosanoids produced

3. Modulation of gene expression

4. Change in lipid-based inflammatory mediators

5. Alteration in energy (calorie) consumption and metabolism

6. Production of epoxides and peroxides

7. Change in membrane fluidity

8. Secretion of mammogenic hormones

9. Alteration in intercellular interaction

of dietary n-6 polyunsaturated fat[12] whereas natural killer cell activity against tumor targets was not affected by the fat in the diet.[13] Because the macrophage can be pivotal in a number of immune as well as nonspecific functions, we have focused on those cells. For this review we will first discuss very briefly how fatty acids may be metabolized, particularly with reference to the macrophage. Second, we will investigate how fatty acids can alter macrophage function with particular emphasis on tumoricidal activities, including the release of soluble tumoricidal mediators such as tumor necrosis factor-alpha (TNFα). Finally, we will discuss some preliminary evidence with respect to mechanisms by which dietary fats alter those macrophage activities. Hopefully, an understanding of how lipids regulate macrophage tumoricidal function might in turn lead to insights of how dietary fat may be manipulated to affect breast tumor regression.

II. Metabolism of Dietary Fat by Cells of the Immune System

Most dietary fats form as triacylglycerols. They vary greatly with respect to the amount and the location of the fatty acid groups esterified to the glycerol backbone. Those fatty acids may be released during digestion and absorbed as free fatty acids which can be reacylated before entry into circulation as chylomicrons. Once these lipids arrive at organs, particularly the liver, they serve as a source of fatty acids for lipoproteins. To provide a source of energy, those fatty acids are metabolized by β-oxidation. Additionally, fatty acids from dietary sources may also be incorporated into phospholipids which form part of the structure of cellular membranes. Thus, macrophages may incorporate dietary derived fatty acids into their membranes.[14-16] The site of alteration may be important because membranes appear to be able to carry out numerous biological functions. Those functions are accomplished within a wide range of fatty acid compositions.

Dietary polyunsaturated fatty acids (PUF) may serve as important substrates which may be then metabolized to form numerous biologically active compounds. Select fatty acids such as linoleic (18:2n-6), oleic (18:1n-9) and α-linolenic (18:3n-3) may be metabolized by alternating steps of desaturation (Δ6, Δ5 desaturase) and elongation to form arachidonic (20:4n-6), eicosatrienoic (20:3n-9), and eicosapentaenoic (20:5n-3) fatty acids. The Δ6 desaturase is rate limiting and is absent in macrophages[17] whereas Δ5 desaturase is functional.[18] Essential fatty acid deficiency as well as other fatty acids may influence desaturase activity.[19] For example, n-3 fatty acids such as those found in abundance in fish oils can depress desaturase activity which can be accentuated when dietary linoleic acid levels are limited.[20] Competition between desaturases and acyl transferases may influence the amount of fatty acid which is available to the macrophage.

Arachidonic acid, found in the cell membrane, can be released by a number of different stimuli and then either re-esterified or metabolized to various eicosanoids, 20-carbon fatty acid metabolites, such as prostaglandins (PG), leukotrienes (LT) and hydroxy fatty acids. Numerous studies have shown that dietary 18:2n-6 can affect eicosanoid levels.[21,22] Arachidonic acid can be released from the cell membrane by the action of phospholipases (PL), and metabolized by either the cyclooxygenase or lipoxygenase pathway depending on how the macrophage was stimulated. Action by the cyclooxygenase enzyme leads to the formation of PGD_2, PGE_2, PGF_2, thromboxane (TX) A_2 and prostacyclin (PGI_2). Long chain n-3 fatty acids such as those found in fish oils competitively reduce eicosanoid synthesis.[23] The other metabolic pathway involves the 5-, 12-, or 15-lipoxygenase enzymes. The resulting hydroperoxy fatty acids participate in the formation of LT or hydroxy eicosatetraenoic acids (HETE). The role of PG in modulation of macrophage function has been studied much more extensively than LT or HETE derivatives.

Alteration of dietary fat can result in a change in the fatty acid composition of cells of the immune system including lymphocytes and macrophages. Those cells may synthesize nonessential fatty acids but must obtain essential fatty acids, including arachidonic, from the blood plasma. Numerous reports have described alterations of macrophage phospholipids associated with dietary alteration involving vegetable and fish oils.[14-16,24] It has been established that macrophage phospholipids are not resistant to change and that phospholipids may be altered to reflect the fatty acid composition of the diet. However, the issue of why certain long-chain fatty acids are preferentially acylated to a particular phospholipid species may be complex and is not completely understood. Thus, although it is possible to alter macrophage phospholipid composition, acyl turnover in macrophage phospholipids through hydrolysis and reacylation is tightly controlled.[25]

III. Macrophage Function and Activation

Macrophages play an important role in the maintenance of homeostasis. In particular they assume a pivotal role in both inflammatory and immune responses, having multiple functions which could be both afferent and efferent in nature (Table 2). For this purpose, highly differentiated or activated macrophages are capable of killing tumor cells. Many of these important functions are not constitutively expressed, but they are induced in response

Table 2. The Role of Macrophages in Maintenance of Homeostasis

1. Removal of debris, effete cells, and serum proteins

2. Host defense against microbial invasion

3. Host defense against neoplasia

4. Response to injury, inflammation

5. Produce factors that modify immune system

6. Can be modified by an array of factors

to precise signals in the microenvironment. Moreover, the capacity to express such functions is usually limited, since overexpression may be detrimental to the host. An understanding of the macrophage and regulation of its many functions may depend upon an understanding of substrate utilization and autoregulatory products. Dietary fatty acids appear to be one such substrate where metabolism can result in the formation of important regulatory substances.

Macrophage activation may be initiated *in vivo*[26,27] or *in vitro* by exposure to various signals. For this the macrophage passes through distinct stages in order to acquire full tumoricidal capability (Figure 1). Through their process of maturation, macrophages normally acquire several functions such as endocytosis and secretion.[28] Although the biochemical basis for activation is not completely understood, it probably involves increases or decreases in membrane-bound and secreted proteins as well as alteration in signal transduction events. The initial stage of macrophage activation is represented by the resident macrophage such as those taken from the peritoneum. Macrophages obtained from sites of inflammation are predominantly newly immigrated from peripheral blood and thus are capable of responding to various inductive signals. These macrophages respond to interferon-gamma (IFNγ) in such a way that they are primed for cytolysis. They become fully capable of killing tumors when triggered by a second signal such as low doses of lipopolysaccharide (LPS).[29,30] Regulation for tumor cell killing occurs over a 12-24 hr period. IFNγ treatment also leads to modulation of protein kinase C (PKC) which results in increased catalytic efficiency of the enzyme when macrophages are exposed to the appropriate stimulus.[31] Previous studies have suggested a linear progression model for macrophage

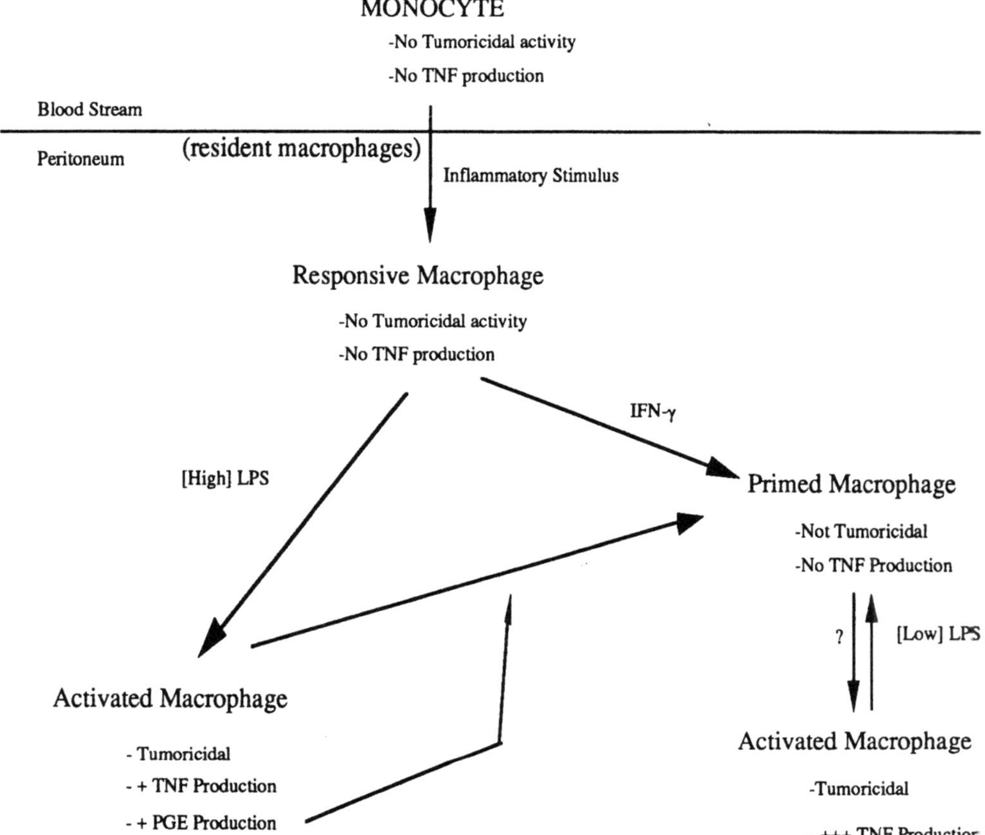

Figure 1. Select events in the pathway of macrophage activation.

activation. However, with the discovery of other agents such as interleukin-4, granulo-cyte/macrophage-colony stimulating factor (GM-CSF) and colony stimulating factor-1, *in vivo* macrophage activation may be more accurately represented by a network model involving several stimulating and suppressive agents. IFNγ has also been shown to alter Ca^{2+} flux. The effects of LPS include the rapid hydrolysis of polyphosphoinositides and the subsequent generation of diacylglycerol (DAG), fluxation of Ca^{2+}, phosphorylation and myristoylation of specific substrates. LPS-induced myristoylation appears to be important to the priming of macrophages for the subsequent release of arachidonic acid. A number of those arachidonic acid metabolites may have an important role in regulation or activation. Thus, macrophage activation may be viewed as a network of different functional states with each having the potential for modulation to another state, depending upon what inductive or suppressive signals are given.[28,32]

Both human and murine macrophages can selectively lyse neoplastic cells while leaving non-neoplastic cells unharmed. Such capacity is not constitutively expressed but is under rigorous regulation. In this process there appear to be several steps. First, there is a close physical interaction between the macrophage and target. Next, the macrophage secretes one or more potentially lytic substances in the diffusion-limited space between effector and target cell.[33] The most likely mediators are reactive oxygen intermediates like H_2O_2 or superoxide anion, cytolytic proteases, TNFα and reactive nitrogen intermediates.[34-36] It should be noted that there are degrees of target sensitivity for each of those lytic agents. Moreover, some of those functions appear to be affected by changes in dietary fat whereas others are not.[37,38] Thus, the ability of the macrophage to destroy a given tumor cell depends upon the regulation of various lytic mechanisms.

IV. Dietary Fat Modulation of Macrophage Function

Fatty acids delivered both *in vitro* and *in vivo* may be rapidly incorporated into macrophages and selectively alter the fatty acyl composition of the macrophage lipids. A number of functional activities can also be modified by delivery of fatty acids *in vitro* such as pinocytosis and phagocytosis. Dietary fats, such as those high in n-3 fatty acids, have been shown to prolong bleeding times, protect against thrombosis, rheumatoid and collagen-induced arthritis as well as glomerular nephritis.[39-44] Thus, it appears appropriate to determine the relationship of dietary fat to macrophage tumoricidal capacity.

A. Alteration in Tumoricidal Capacity

Dietary fatty acids, particularly those found in marine fish oils, may be very effective in altering macrophage functions, particularly those associated with an inflammatory response. To assess how dietary fish oil, which is rich in n-3 fatty acids, compared to vegetable oils, rich in n-6 fatty acids, could differentially affect macrophage tumoricidal capacity, mice were fed isocaloric diets which contained 10% by weight of fat from menhaden fish oil (MFO) or safflower oil (SAF).[45] No differences between the diets with respect to the total number of peritoneal exudate cells or the percentage of macrophages were observed. Macrophages from mice fed MFO killed fewer tumor target cells upon activation with IFNγ and low doses of LPS than macrophages from mice fed the SAF diet (Table 3). Possible explanations for the decreased activity are that the macrophages from animals fed MFO were defective in the actual cytolytic mechanism, or that they did not respond to the priming agent, IFNγ, or the activating agent, LPS as well as macrophages from mice fed SAF. To distinguish between those possibilities, macrophages were primed pharmacologically with phorbol myristate acetate. That agent mimics IFNγ with respect to signal transduction mechanisms, such as the induction of protein kinase C activity. All groups of

Table 3. Effect of Diet on the Requirement of IFNγ for *in vitro* Tumoricidal Activation

	IFNγ(U/ml)	
	2.5	25
SAF	50 ± 2[a]	64 ± 2
MFO	19 ± 2	65 ± 3

[a] Values represent mean ± S.E.M. percent specific lysis of P815 tumor cells exposed 18 hr to macrophages activated for 4 hr with IFNγ and 10 ng/ml LPS.

macrophages were equally competent for tumoricidal activity and equally competent when additional LPS was added. The finding that cytolytic capacity was enhanced equally suggests that dietary manipulation did not alter LPS responsiveness. Observations that priming with higher concentrations of IFNγ restored the partial defect in activation indicates that macrophages from MFO-fed mice were hyporesponsive to IFNγ (Table 3). To assess the putative hyporesponsiveness to IFNγ of macrophages from mice fed MFO, two additional processes sensitive to IFNγ were measured (Table 4). Although the two dietary groups of macrophages produced similar quantities of peroxide at low concentrations of IFNγ, with high levels of IFNγ, fish oil macrophages released more peroxide. When macrophages were stimulated for peroxide production with unopsonized zymosan, macrophages from mice fed SAF produced higher levels of peroxide compared with macrophages from mice fed MFO.[46] In contrast, no differences were observed between the two diets when macrophages were stimulated and Ia expression measured. Thus, the differences between macrophages from mice fed MFO and SAF may be through an alteration of IFNγ-induced functions such as tumoricidal activity and peroxide production.[46]

The putative anti-inflammatory effect of dietary MFO may be due to alteration in eicosanoid production. Thus, autoregulation of macrophage-produced PGE_2 was examined by stimulation with high levels of LPS for 24 hrs. In those experiments, macrophages from mice fed SAF had a lower cytolytic capacity than macrophages from mice fed MFO. Addition of a PG inhibitor, indomethacin, resulted in enhanced levels of tumor cell kill by macrophages from mice fed SAF and no alteration in cytolytic capacity in macrophages from mice fed fish oil (Figure 2). When PGE levels were assessed, macrophages from SAF fed mice produced three times more PGE than did macrophages from mice fed MFO. These results indicate that dietary fats can alter tumoricidal capacity of macrophages, possibly involving mechanisms both dependent and independent of changes in eicosanoid synthesis. This may be important as any change in regulation of macrophage activation may be critical for multiple *in vivo* functions such as tumoricidal activity.

B. Alteration of Cytolytic Factor Production

Tumor necrosis factor-alpha (TNFα), also known as cachectin, was originally described as a serum factor, derived from animals following a bacterial challenge, causing necrosis of certain transplanted tumors.[47] TNFα is an important inflammatory mediator that may have an effect on the function of numerous tissues and organs (Table 5). Circulating TNFα is usually associated with a significant pathological change, possibly leading to mortality.[48,49] TNFα has a multitude of biological activities, including effects on vascular endothelium, hematopoietic elements, adipose tissue, muscle, the gastrointestinal tract, brain, skin and bone. Thus TNFα synthesis must be stringently controlled. A principal source of TNFα is

Table 4. Effect of Dietary Fat on Zymosan-Stimulated and IFNγ-primed Peroxide Production

	Diet	
Stimulating agent	SAF	MFO
	Peroxide (nM)	
Zymosan (µg/ml)		
0	0	0
10	25 ± 2	0
50	60 ± 5	40 ± 5
100	140 ± 8	50 ± 4
200	280 ± 8	140 ± 2
IFNγ (U/ml)+PMA (100 nM)		
0	38 ± 2	38 ± 2
0.1	46 ± 1	47 ± 5
1	63 ± 3	56 ± 2
10	68 ± 4	73 ± 3
100	69 ± 2	93 ± 2

Table 5. Effects of TNFα/Cachectin in Select Tissues.

Tissue	Effect
Muscle	Stimulates glucose uptake, enhances protein breakdown; accelerates glycogenesis
Adipose	Suppresses LPL; promotes free fatty acid and triglyceride efflux; inhibits lipogenic enzyme biosynthesis
Liver	Suppresses albumin synthesis; stimulates lipogenesis; induces acute phase protein biosynthesis
Brain	Mediates fever by inducing PG synthesis; induces anorexia; increases pituitary release of ACTH
Endothelium	Stimulates biosynthesis of IL-1 and PAF; induces HLA and ICAM's; increases procoagulant activity
Connective tissue	Stimulates IL-6 biosynthesis; induces PG and collagenases; mediates bone resorption and calcium release
Leukocytes	Induces interleukin-1, TNFα, GM-colony stimulating factor, platelet derived growth factor, and transforming growth factor-β; enhances production of reactive oxygen intermediates, mediates increased toxicity against parasites and fungi

the macrophage which can also secrete large quantities of eicosanoids.[50] The effects of dietary fat on monocyte/macrophage production of TNFα have been studied in mice, rats and humans.[38,51-53] Although regulation of TNFα synthesis can occur at multiple levels, production may be inhibited by PGE$_2$.[54] While several previous studies have demonstrated that inclusion of fish oils in the diet can reduce PGE$_2$ production by macrophages, the mechanisms by which fish oil-containing diets may alter specific aspects of macrophages and TNFα production remain largely unknown. To assess how dietary fatty acids may regulate TNFα production, mice were fed MFO as a source of n-3 fatty acids or SAF as a control and source of n-6 fatty acids. Although macrophages from both groups of mice produced

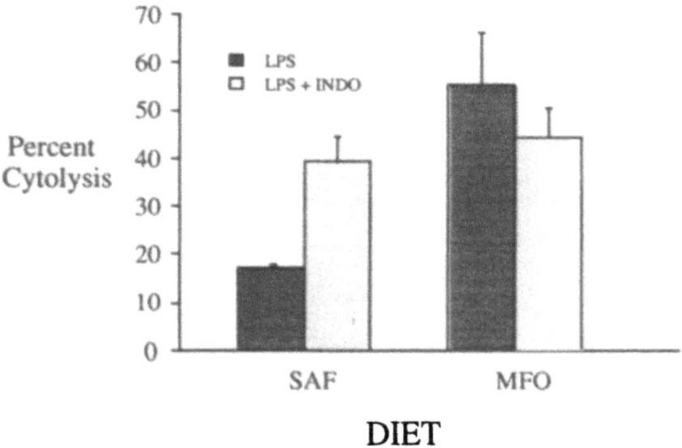

Figure 2. Effect of diatary fat on macrophage activation by LPS for cytolytic activity. Macrophage monolayers from mice fed safflower oil (SAF) or menhaden fish oil (MFO) diets were incubated with 100 ng/ml LPS with or without 10^{-6} M indomethacin (INDO) for 24 hrs. Cytolysis was measured 16 hrs after addition of radiolabeled tumor targets.

TNFα which peaked at 6-8 hrs after LPS stimulation, macrophages from mice fed MFO produced significantly more TNFα at 24 hrs than macrophages from mice fed SAF (Table 6). The relatively high levels of TNFα found in MFO macrophage cultures at 24 hrs did not appear to be due to continuous production of bioactive TNFα. Production was self-limiting and basically complete by about 8 hrs after LPS exposure. The down-regulation of TNFα production may be due to an alteration in transcription.[55] However, there were no differences in mRNA levels for both macrophage populations after LPS stimulation. Lack of TNFα production after 8 hrs and very little TNFα transcription at 24 hrs indicates that suppression was more complex than just termination of transcription. In addition, macrophages from mice fed SAF were able to remove secreted TNFα, whereas macrophages from mice fed MFO were apparently unable to do the same. Also, addition of indomethacin

Table 6. Effect of Dietary Fat on the Kinetics of LPS-Stimulated TNFα Production

	Diet	
Time (hr)	SAF	MFO
	TNF (units)	
0	0	0
1	20 ± 1	20 ± 1
6	200 ± 10	295 ± 15
24	40 ± 1	280 ± 15

caused a significant increase in TNFα levels after 24 hrs of LPS stimulation of macrophages from mice fed SAF whereas TNFα production by macrophages from mice fed fish oil was unaltered.[38] To examine the possibility that PGE$_2$ may be associated with the removal of TNFα that was secreted into the culture medium, macrophages from mice fed both diets were treated with LPS, indomethacin to block endogenous PGE$_2$ production and various concentrations of exogenous PGE$_2$ (Table 7). The amount of PGE$_2$ required to inhibit TNFα at 8 hrs was similar for both populations of macrophages. At 24 hrs, however, macrophages from mice fed SAF required one log less PGE$_2$ than macrophages from those fed MFO for suppression of TNFα. This suggests that at least two elements are associated with PGE$_2$-mediated regulation of TNFα production. A concentration of PGE$_2$ not altering synthesis of TNFα at 8 hrs of culture appeared to indicate the clearance of secreted, soluble TNFα.

V. Possible Mechanisms of Dietary Fat Induced Alteration of Macrophage Function

There are at least six possible mechanisms by which dietary fat may influence how the macrophage can interact and kill tumor cells (Table 8). Experimental evidence has been published for some of those while others remain speculative. As demonstrated above, dietary fat may alter either the type or quantity of eicosanoids produced, or both. With respect to cells of the immune system, macrophages are the main producers of those compounds. In addition to important proinflammatory characteristics, eicosanoids can also suppress numerous leukocyte functions. However, the exact concentrations of eicosanoids often govern their effect. Besides their effect on lymphocytes, macrophage-produced eicosanoids may play an important role in autoregulation of macrophage functions such as down-regulation of cytolytic capacity. In addition, 5-lipoxygenase products such as LTB$_4$ and LTC$_4$ are produced by macrophages after stimulation and may have an effect on macrophage function such as the ability to lyse tumor targets.[56] Moreover, changes in the type of fatty acids that are presented in the diet may affect the relative amounts of both PG and LT produced. Thus, changes in the eicosanoids produced remain a logical explanation.

Another possible mechanism is that dietary fat may cause an alteration in signal transduction. In macrophage activation, LPS may activate phosphatidylinositol turnover with a subsequent rapid rise in intracellular Ca^{2+}.[57] The action of phospholipase C on phosphatidylinositol leads to the generation of diacylglycerol (Figure 3). Phospholipid derivatives such as diacylglycerol are important components in the activation of protein kinase C (PKC). LPS is also known to induce protein phosphorylation by PKC.[57] Since it is possible to alter the acyl composition of phospholipids by dietary fat, it is possible that

Table 7. Effect of Dietary Fat on the Inhibition of TNFα Production by PGE$_2$

	Diet	
PGE$_2$(nM)	SAF	MFO
	Percent inhibition	
0	0	0
1	19	0
2	62	5
10	90	22
50	90	35

Table 8. Possible Mechanisms by Which Dietary Fat Could Modify Macrophage Function

1. Alteration of the type and quantity of eicosanoids produced

2. Change in membrane fludity

3. Alteration in intercellular interaction

4. Modulation of signal transduction

5. Changes in lipid-based inflammatory mediators

6. Alteration in gene regulation

dietary fat may alter PKC activity. Accordingly, we assessed the influence of MFO and SAF on IFNγ-enhanced PKC activity. With 25 U/ml of IFNγ, total PKC activity was much greater in macrophages from mice fed SAF compared to those fed MFO (Figure 4A). Further experiments were performed to test the effect of diet on the translocation of PKC from the cytosol to the cell membranes in response to IFNγ. Translocation of PKC from the cytosol to the plasma membrane appeared to be required for its activity (Figure 4B). Without IFNγ treatment, most of the PKC activity (70-80%) was localized in the cytosol. However, with IFNγ treatment, most of the PKC activity was in the membrane fraction. IFNγ stimulated a 50% loss of cytosolic PKC activity within 60 sec in macrophages of mice fed SAF, whereas the 50% loss was seen in macrophages from mice fed MFO after 8 min. After 30 min, levels of PKC were higher in the membranes of macrophages from mice fed SAF compared to those fed MFO. These results indicate that macrophages from mice fed MFO versus SAF have an altered response to IFNγ for enhancement of PKC activity. In addition, hyporesponsiveness to IFNγ was observed in macrophages from mice fed MFO with respect to priming for cytolysis and may be due to effects on PKC.

A third possible mechanism may be that dietary fat is associated with an alteration in gene regulation. Recently we have identified an antiproliferative gene whose expression was enhanced two-fold after treatment with PGE$_2$.[58] That enhancing effect was also exerted by PGE$_1$ and platelet activation factor, but not LTB$_4$, LTC$_4$ or LPS. This antiproliferative gene may play a role in PGE_2-mediated inhibition of macrophage proliferation. It may also be possible to alter its expression by altering dietary fat and the endogenous production of PGE_2.

VI. Summary

It has been known for at least 20 years that fatty acids can alter immune functions *in vitro*. More recently we have begun to understand the role that dietary fats play in immunity

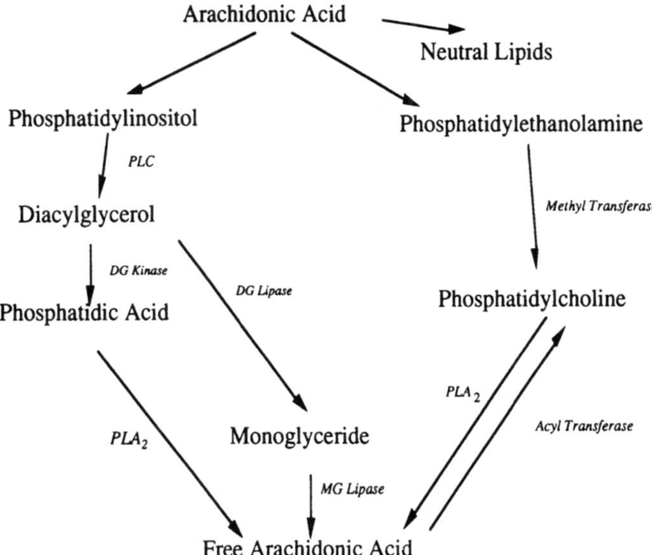

Figure 3. Pathway by which arachidonic acid may be incorporated into or removed from phospholipids.

and specifically how they may alter macrophage function. In the future it will be important not simply to redefine that fatty acids can alter select macrophage functions but to understand the mechanisms by which that occurs. Whether the same or different mechanisms are operational for those functions that are altered by dietary fat remains to be determined. Nevertheless, tumoricidal responses can be modified depending on the fatty acids in the diet. Hopefully, these recent observations will expand our understanding of how lipids regulate macrophage tumoricidal function and thus, might lead to new insights of how dietary fat may be manipulated to affect breast tumor regression.

Acknowledgments

The authors' work cited herein was supported by grant CA 47050 from the National Cancer Institute, and by a grant from the National Dairy Board, administered by the National Dairy Council.

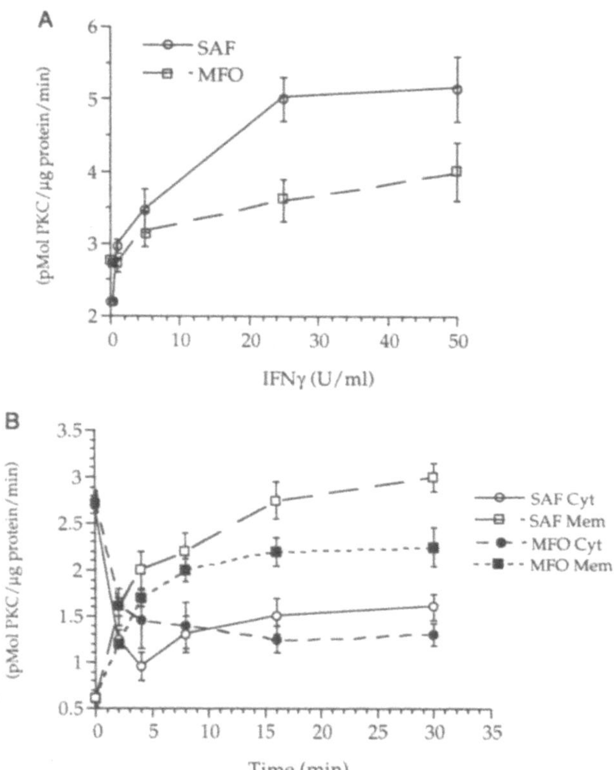

Figure 4. Effect of dietary fat on protein kinase C (PKC) activity. A, total PKC, B, translocation of PKC from the cytosol (Cyt) to the cell membrane (Mem).

References

1. Prentice, R.L., Dietary fat reduction as a hypothesis for the prevention of postmenopausal breast cancer, and a discussion of hypothesis testing research strategies, *in: Diet and Breast Cancer*, E.K. Weisburger, ed., Plenum Press, New York (1994).
2. Willett, W.C., D.J. Hunter, M.J. Stampfer, G. Colditz, J.E. Manson, D. Spiegelman, B. Rosner, C.H. Hennekens, and F.E. Speizer, Dietary fat and fiber in relation to risk of breast cancer. An 8-year follow-up, *JAMA* 268:2037 (1992).
3. Harris, J.R., M.E. Lippman, U. Veronesi, and W. Willett, Breast cancer (1), *N. Engl. J. Med.* 327:319 (1992).
4. Harris, J.R., M.E. Lippman, U. Veronesi, and W. Willett, Breast cancer (2), *N. Engl. J. Med.* 327:390(1992).
5. Harris, J.R., M.E. Lippman, U. Veronesi, and W. Willett, Breast cancer (3), *N. Engl. J. Med.* 327:473 (1992).
6. Welsch, C.W., Enhancement of mammary tumorigenesis by dietary fat: review of potential mechanisms, *Am. J. Clin. Nutr.* 45:192 (1987).
7. Carroll, K.K. and G.J. Hopkins, Dietary polyunsaturated fat versus saturated fat in relation to mammary carcinogenesis, *Lipids* 14:155 (1979).
8. Hillyard, L.A. and S. Abraham, Effect of dietary polyunsaturated fatty acids on growth of mammary adenocarcinomas in mice and rats, *Cancer Res.* 39:4430 (1979).
9. Erickson, K.L. and N.E. Hubbard, Dietary fat and tumor metastasis, *Nutr. Rev.* 48:6 (1990).
10. Ip, C., C.A. Carter, and M.M. Ip, Requirement of essential fatty acid for mammary tumorigenesis in the rat, *Cancer Res.* 45:1997 (1985).
11. Hubbard, N.E. and K.L. Erickson, Enhancement of metastasis from a transplantable mouse mammary tumor by dietary linoleic acid, *Cancer Res.* 47:6171 (1987).
12. Erickson, K.L. and I.K. Thomas, Susceptibility of mammary tumor cells to complement-mediated cytolysis after *in vitro* or *in vivo* fatty acid manipulation, *J. Natl. Cancer Inst.* 75:333 (1985).
13. Erickson, K.L. and L.A. Schumacher, Lack of an influence of dietary fat on murine natural killer cell activity, *J. Nutr.* 119:1311 (1989).
14. Lokesh, B.R., H.L. Hsieh, and J.E. Kinsella, Peritoneal macrophages from mice fed dietary (n-3) polyunsaturated fatty acids secrete low levels of prostaglandins, *J. Nutr.* 116:2547 (1986).
15. Lokesh, B.R., J.M. Black, J.B. German, and J.E. Kinsella, Docosahexaenoic acid and other dietary polyunsaturated fatty acids suppress leukotriene synthesis by mouse peritoneal macrophages, *Lipids* 23:968 (1988).
16. Broughton, K.S., J. Whelan, I. Hardardottir, and J.E. Kinsella, Effect of increasing the dietary (n-3) to (n-6) polyunsaturated fatty acid ratio on murine liver and peritoneal cell fatty acids and eicosanoid formation, *J. Nutr.* 121:155 (1991).
17. Chapkin, R.S., S.D. Somers, and K.L. Erickson, Dietary manipulation of macrophage phospholipid classes: selective increase of dihomogammalinolenic acid, *Lipids* 23:766 (1988).
18. Chapkin, R.S., N.E. Hubbard, D.K. Buckman, and K.L. Erickson, Linoleic acid metabolism in metastatic and nonmetastatic murine mammary tumor cells, *Cancer Res.* 49:4724 (1989).
19. Brenner, R.R., H. Garda, H. DeGremes, I. Dumm, and H. Pezzano, Early effects of EFA deficiency on the structure and enzymatic activity of rat liver microsomes, *Prog. Lipid Res.* 20:315 (1982).
20. DeSchrijuer, R. and O.S. Privett, Effects of dietary long chain fatty acids on the biosynthesis of unsaturated fatty acids in the rat, *J. Nutr.* 112:619 (1982).
21. Adams, O., G. Wolfram, and N. Zollner, Prostaglandin formation in man during intake of different amounts of linoleic acids in diets, *Ann. Nutr. Metab.* 26:315 (1982).
22. Dupont, J., Essential fatty acids and prostaglandins, *Prev. Med.* 16:485 (1987).
23. Lands, W.E.M., Renewed questions about polyunsaturated fatty acids, *Nutr. Rev.* 44:189 (1986).

24. Lokesh, B.R., J.M. Black, and J.E. Kinsella, The suppression of eicosanoid synthesis by peritoneal macrophages is influenced by the ratio of dietary docosahexaenoic acid to linoleic acid, *Lipids* 24:589 (1989).

25. Kuwae, T., P.C. Schmid, S.B. Johnson, and H.H.O. Schmid, Differential turnover of phospholipid acyl groups in mouse peritoneal macrophages, *J. Biol. Chem.* 265:5002 (1990).

26. Hibbs, J.B., H.A. Chapman, and J.B. Weinberg, The macrophage as an antineoplastic surveillance cell: biological perspectives, *J. Reticuloendothel. Soc.* 24:549 (1978).

27. Cleveland, R.P., M.S. Meltzer, and B. Zbar, Tumor cytotoxicity *in vitro* by macrophages from mice infected with *Mycobacterium bovis* strain BCG, *J. Natl. Cancer Inst.* 52:1887 (1974).

28. Adams, D.O. and T.J. Koerner, Gene regulation in macrophage development and activation, *Year. Immunol.* 4:159 (1989).

29. Pace, J.L., S.W. Russell, B.A. Torres, H.M. Johnson, and P.W. Gray, Recombinant mouse γ interferon induces the priming step in macrophage activation for tumor cell killing, *J. Immunol.* 130:2011 (1983).

30. Schreiber, R.D., J.L. Pace, S.W. Russell, A. Altman, and D.H. Katz, Macrophage-activating factor produced by a T cell hybridoma: physiochemical and biosynthetic resemblance to gamma-interferon, *J. Immunol.* 131:826 (1983).

31. Becton, D.L., D.O. Adams, and T.A. Hamilton, Characterization of protein kinase C activity in interferon gamma treated murine peritoneal macrophages, *J. Cell. Physiol.* 125:485 (1985).

32. Adams, D.O. and T.A. Hamilton, The cell biology of macrophage activation, *Annu. Rev. Immunol.* 2:283 (1984).

33. Nathan, C.F., Secretory products in cytotoxicity, in: *Biological Response Mediators and Modulators*, J.T. August, ed., Academic Press, New York (1983).

34. Marino, P.A. and D.O. Adams, Interaction of Bacillus Calmette—Guerin-activated macrophages and neoplastic cells *in vitro*. I. Conditions of binding and its selectivity, *Cell. Immunol.* 54:11 (1980).

35. Adams, D.O., Effector mechanisms of cytolytically activated macrophages. I. Secretion of neutral proteases and effect of protease inhibitors, *J. Immunol.* 124:286 (1980).

36. Adams, D.O., K.J. Kao, R. Farb, and S.V. Pizzo, Effector mechanisms of cytolytically activated macrophages. II. Secretion of a cytolytic factor by activated macrophages and its relationship to secreted neutral proteases, *J. Immunol.* 124:293 (1980).

37. Hubbard, N.E., D. Lim, S.D. Somers, and K.L. Erickson, Effects of *in vitro* exposure to arachidonic acid on TNF-α production by murine peritoneal macrophages, *J. Leukoc. Biol.* 54:105 (1993).

38. Somers, S.D. and K.L. Erickson, Alteration of tumor necrosis factor-α production by macrophages from mice fed diets high in eicosapentaenoic and docosahexaenoic fatty acids, *Cell. Immunol.* 153:287 (1994).

39. Bang, H.O., J. Dyerberg, and N. Hjorne, The composition of food consumed by Greenland Eskimos, *Acta. Med. Scand.* 200:69 (1976).

40. Whitaker, M.O., A. Wyche, F. Fitzpatrick, H. Sprecher, and P. Needleman, Triene prostaglandins: prostaglandin D3 and icosapentaenoic acid as potential antithrombotic substances, *Proc. Natl. Acad. Sci. USA* 76:5919 (1979).

41. Dyerberg, J., H.O. Bang, E. Stoffersen, S. Moncada, and J.R. Vane, Eicosapentaenoic acid and prevention of thrombosis and atherosclerosis?, *Lancet* ii:117 (1978).

42. Kremer, J.M., J. Bigauoette, A.V. Michalek, M.A. Timchalk, L. Lininger, R.I. Rynes, C. Huyck, J. Zieminski, and L.E. Bartholomew, Effects of manipulation of dietary fatty acids on clinical manifestations of rheumatoid arthritis, *Lancet* i:184 (1985).

43. Prickett, J.D., D.E. Trentham, and D.R. Robinson, Dietary fish oil augments the induction of arthritis in rats immunized with type II collagen, *J. Immunol.* 132:725 (1984).

44. Leslie, C.A., W.A. Gonnerman, M.D. Ullman, K.C. Hayes, C. Franzblau, and E.S. Cathcart, Dietary fish oil modulates macrophage fatty acids and decreases arthritis susceptibility in mice, *J. Exp. Med.* 162:1336 (1985).

45. Somers, S.D., R.S. Chapkin, and K.L. Erickson, Alteration of *in vitro* murine peritoneal macrophage function by dietary enrichment with eicosapentaenoic and docosahexaenoic acids in menhaden fish oil, *Cell. Immunol.* 123:201 (1989).

46. Hubbard, N.E., S.D. Somers, and K.L. Erickson, Effect of dietary fish oil on development and selected functions of murine inflammatory macrophages, *J. Leukoc. Biol.* 49:592 (1991).

47. Carswell, E.A., L.J. Old, R.L. Kassel, S. Green, N. Fiore, and B. Williamson, An endotoxin-induced serum factor that causes necrosis of tumors, *Proc. Natl. Acad. Sci. USA* 72:3666 (1975).

48. Tracey, K.J., B. Beutler, S.F. Lowry, J. Merryweather, S. Wolpe, I.W. Milsark, R.J. Hariri, T.J. Fahey III, A. Zentella, J.D. Albert, G.T. Shires, and A. Cerami, Shock and tissue injury induced by recombinant human cachectin, *Science* 234:470 (1986).

49. Tracey, K.J., H. Wei, K.R. Manogue, Y. Fong, D.G. Hesse, H.T. Nguyen, G.C. Kuo, B. Beutler, R.S. Cotran, and A. Cerami, Cachectin/tumor necrosis factor induces cachexia, anemia, and inflammation, *J. Exp. Med.* 167:1211 (1988).

50. Beutler, B. and A. Cerami, Tumor necrosis, cachexia, shock, and inflammation: a common mediator, *Annu. Rev. Biochem.* 57:505 (1988).

51. Hardardottir, I. and J.E. Kinsella, Tumor necrosis factor production by murine resident peritoneal macrophages is enhanced by dietary n-3 polyunsaturated fatty acids, *Biochim. Biophys. Acta* 1095:187 (1991).

52. Vilcek, J. and T.H. Lee, Tumor necrosis factor, *J. Biol. Chem.* 266:7313 (1991).

53. Endres, S., R. Ghorbani, V.E. Kelley, K. Georgilis, G. Lonnemann, J. Vandermeer, J.G. Cannon, T.S. Rogers, M.S. Klempner, P.C. Weber, E.J. Schaefer, S.M. Wolff, and C.A. Dinarello, The effect of dietary supplementation with n-3 polyunsaturated fatty acids on the synthesis of interleukin-1 and tumor necrosis factor by mononuclear cells, *N. Engl. J. Med.* 320:265 (1989).

54. Kunkel, S.L., M. Spengler, G. Kwon, M.A. May, and D.G. Remick, Production and regulation of tumor necrosis factor alpha, *Methods Achiev. Exp. Pathol.* 13:240 (1988).

55. Kunkel, S.L., M. Spengler, M.A. May, R. Spengler, J. Larrick, and D. Remick, Prostaglandin E_2 regulates macrophage-derived tumor necrosis factor gene expression, *J. Biol. Chem.* 263:5380 (1988).

56. Braun, D.P., M.-C. Ahn, J.E. Harris, E. Chu, L. Casey, G. Wilbanks, and K.P. Siziopikou, Sensitivity of tumoricidal function in macrophages from different anatomical sites of cancer patients to modulation of arachidonic acid metabolism, *Cancer Res.* 53:3362 (1993).

57. Prpic, V., J.E. Weiel, S.D. Somers, J. DiGuiseppi, S.L. Gonias, S.V. Pizzo, T.A. Hamilton, B. Herman, and D.O. Adams, Effects of bacterial lipopolysaccharide on the hydrolysis of phosphatidylinositol-4,5 biphosphate in murine peritoneal macrophages, *J. Immunol.* 139:526 (1987).

58. Suk, K. and K.L. Erickson, Enhancement of BTG 1 gene expression by prostaglandin E_2 in macrophages, Submitted for publication (1994).

Chapter 8

Dietary Fatty Acids and Human Breast Cancer Cell Growth, Invasion, and Metastasis

DAVID P. ROSE, JEANNE M. CONNOLLY and XIN-HUA LIU

I. Introduction

There is considerable interest in the utilization of a low-fat dietary intervention to reduce the risk of recurrence in postsurgical breast cancer patients with potentially curable disease.[1-5] This concept developed initially from epidemiological and clinical studies, reviewed elsewhere,[5,6] but it gained support from investigations utilizing experimental animal models. Prominent among these was the work of Erickson and colleagues,[7] who showed that feeding a high-fat diet rich in linoleic acid (LA), an omega-6 polyunsaturated fatty acid, enhanced metastasis of a mouse mammary tumor cell line, compared with a diet containing the same *quantity* of fat, but low in LA content. Similar results were obtained by Katz and Boylan[8] with a transplantable, metastasizing, rat mammary carcinoma, but with diets containing either 20% or 5% (wt/wt) corn oil, a lipid source rich in LA.

Linoleic acid is metabolized by a system of elongase and desaturase enzymes to arachidonic acid. This omega-6 fatty acid is incorporated into cell membrane phospholipids, but it is available for mobilization under the influence of phospholipase A_2, and subsequent enzymatic conversion to eicosanoids (Figure 1). Hubbard *et al.*[9] demonstrated that the stimulatory effect of a high-fat, LA-rich diet on metastasis of mouse mammary carcinoma to the lungs could be suppressed by treatment with indomethacin, a drug which at appropriate concentrations can inhibit both cyclooxygenase and 5-lipoxygenase activities (Figure 1).

II. Effects of Fatty Acids on Cultured Human Breast Cancer Cells

We began our own research program by studying the effects of fatty acids on human breast cancer cell lines cultured *in vitro*. A preliminary study showed that LA stimulates the growth of the estrogen-independent MDA-MB-231 cell line, but that estrogen dependent

---

David P. Rose, Jeanne M. Connolly and Xin-Hua Liu ● Division of Nutrition and Endocrinology, American Health Foundation, Valhalla, New York

Diet and Breast Cancer, Edited under the auspices of the American Institute for Cancer Research, Plenum Press, New York, 1994

MCF-7 cells are less responsive.[10] Later, we reported that whereas LA produces a mitogenic response in MDA-MB-231 breast cancer cells, two omega-3 fatty acids, docosahexaenoic acid (DHA) and eicosapentaenoic acid (EPA), cause suppression of growth.[11] In this same study, we showed that selective inhibitors of leukotriene synthesis, but not of prostaglandin

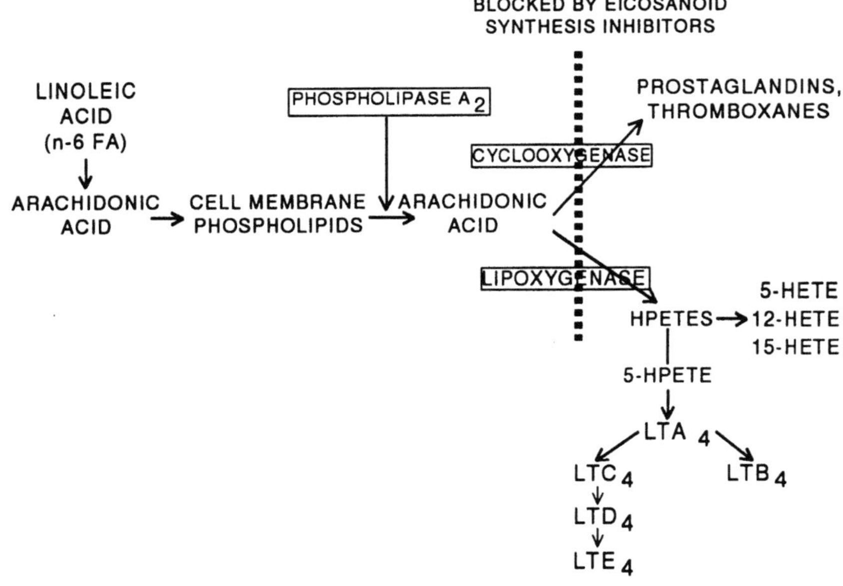

Figure 1. The biosynthesis of eicosanoids from linoleic acid. LT, leukotriene; HPETE, hydroperoxyeicosatetraenoic acid; HETE, hydroxyeicosatetraenoic acid.

synthesis, also suppress growth of these breast cancer cells; indomethacin could block the stimulatory activity of LA, but only at concentrations which inhibit leukotriene as well as prostaglandin production. We interpreted these results as indicating that eicosanoid-responsive breast cancer cells are dependent on one or more of the leukotriene family of eicosanoids, rather than the prostaglandins. A similar conclusion was reached by Buckman et al.[12] when they examined the effects of indomethacin on LA-stimulation of growth of the 4526 mouse mammary tumor cell line.

III. Studies with a Nude Mouse Human Breast Cancer Metastatic Model

The MDA-MB-435 human breast cancer cell line is unusual in that it will reliably metastasize to the regional lymph nodes and lungs when injected into the thoracic mammary fat pads of female athymic nude mice (Figure 2).[13,14] This estrogen-independent cell line exhibits a mitogenic response to LA, which is suppressed completely by concentrations of indomethacin which are capable of blocking leukotriene biosynthesis (Figure 3). The growth rate of the primary tumors which form at the injection site in nude mice can be followed by serial caliper measurements, while the occurrence and extent of metastasis to the lungs can

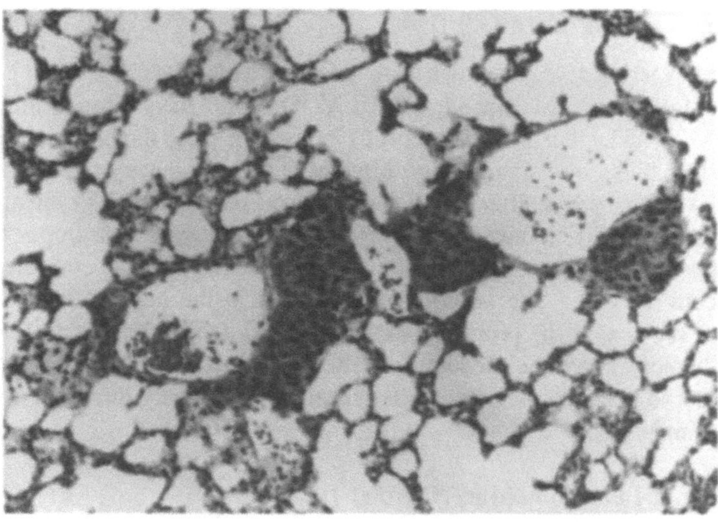

Figure 2. Microscopic MDA-MB-435 human breast cancer cell metastases in lung, originating from a primary tumor mass in a nude mouse thoracic mammary fat pad.

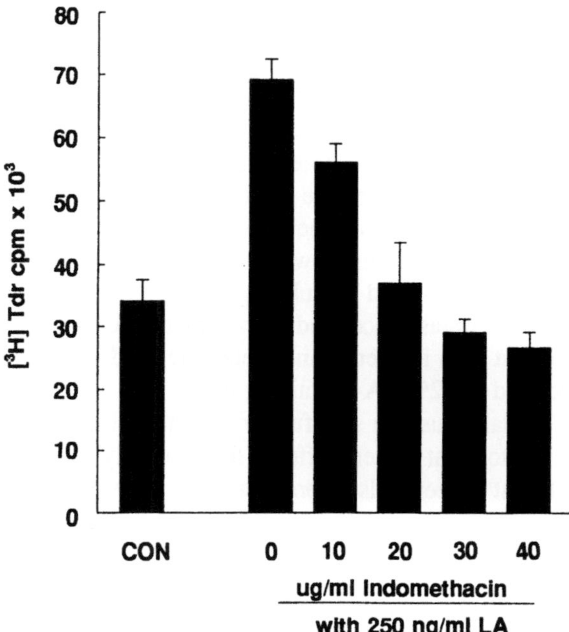

Figure 3. Inhibition of linoleic acid-stimulated [^{3}H]thymidine incorporation into MDA-MB-435 human breast cancer cells *in vitro* by indomethacin. [^{3}H]Thymidine uptake was determined after 3 days exposure of the cultured cells to linoleic acid and the eicosanoid synthesis inhibitor.

be evaluated at necropsy.[15,16] In life, involved axillary lymph nodes may be palpable, and extensive pulmonary involvement is indicated by respiratory distress with tachypnea and cyanosis. We have used this nude mouse model to examine the influence of dietary fat on breast cancer cell growth and metastasis.

A. High- and Low-Fat (Corn Oil) Diets

In our first experiment,[16] nude mice were fed either a high-fat (23% wt/wt corn oil; 12% LA), or a low-fat (5% wt/wt corn oil; 2.7% LA) diet for 7 days prior to the injection of 1×10^6 MDA-MB-435 human breast cancer cells into a right-sided thoracic mammary fat pad. Body weight gains were not significantly different in the two dietary groups, and the diets were continued for 15 weeks, after which the study was terminated. The mammary fat pad tumors grew more rapidly in mice fed the 23% corn oil diet, and at necropsy macroscopic lung metastases were found more frequently and were more extensive compared with mice fed the 5% corn oil diet. The severity of metastasis in the high-fat dietary group was independent of the size of the primary tumor, suggesting the involvement of a mechanism distinct from simply a greater shedding of potentially metastatic tumor cells from large primary cancers; in contrast, metastasis was related to the primary tumor weight in the 5% corn oil-fed animals.

B. Proportion of Linoleic Acid in High-Fat Diets

A recurring concern in discussions on the influence of dietary fat in mammary carcinogenesis is the extent to which the observed promotional effects are due specifically to the lipid, and how much to the associated increase in energy intake. This topic provided the theme of the AICR Second Annual Conference on Nutrition and Cancer.[17]

To explore this issue further in the context of breast cancer metastasis and to examine the influence of dietary LA specifically, we performed an experiment based upon the study by Hubbard and Erickson.[7] Nude mice were assigned to diets which were isocaloric, but contained different mixtures of safflower (LA-rich) and coconut (saturated fatty acid-rich) oils so as to provide 23% (wt/wt) total fat in each case, but with 2, 8 or 12% (wt/wt) LA.[18]

The mammary fat pad tumors grew more rapidly in the mice consuming the 12% and 8% LA-containing diets than in those fed the 2% LA diet, a result which differed from that obtained by Hubbard and Erickson[7] with their 4526 mouse mammary tumor cell line. In their study dietary LA had no effect on growth at the injection site, although it stimulated metastasis, and the same fatty acid did stimulate growth of the cell line *in vitro*.[12]

Grossly visible lung metastases occurred with similar frequency and extent in the 8% and 12% LA-fed mice, but both incidence and metastatic burden were significantly greater than they were in mice fed the 2% LA-containing diet.

This experiment has taken us one step further by establishing unequivocally that diets with the same energy and total fat content differ in their capacity to stimulate the growth and metastasis of human breast cancer cells when a variation exists in their LA content.

C. Dietary Omega-3 Fatty Acid—Fish Oil

It is recognized that dietary omega-3 fatty acids contained in fish oil suppress experimental mammary carcinogenesis in rodents,[19,20] and the progression of transplanted rodent mammary tumors.[21-23] More recently, several investigators have shown that feeding a fish oil-rich diet also inhibits the growth of human breast cancer cells in nude mice.[24-26] However, the transplantable human solid tumors and cell lines utilized in these experiments do not metastasize or do so poorly. Hence we are using the MDA-MB-435 cell line to examine the effect of omega-3 fatty acids on metastasis of human breast cancer cells.

In our initial experiment,[27] female athymic nude mice were injected with MDA-MB-435 cells 7 days after feeding a 23% (wt/wt) corn oil diet. One week later, the mice were assigned to 1 of 3 diets, each of which provided 23% of total fat, but with corn oil and menhaden (fish) oil in the proportions of 18% corn oil to 5% menhaden oil, 11.5% of each oil, or 5% corn oil to 18% menhaden oil. A suppressive effect of a menhaden oil-rich diet on MDA-MB-435 tumor cell growth in the mammary fat pads was observed, which was similar to that reported by others in experiments with different human breast cancer cell lines.[24,26] In addition, when present at a level of 18% (wt/wt), but not 11.5%, the menhaden oil also produced a reduction in the incidence and extent of metastases to the lungs.

These effects of menhaden oil on breast cancer progression and metastasis may result from altered eicosanoid biosynthesis by the tumor cells. Eicosapentaenoic acid, the principal omega-3 fatty acid in menhaden oil, is incorporated into phospholipids at the expense of arachidonic acid.[28] It can inhibit phospholipase A_2 activity[29] and thus suppresses synthesis of eicosanoids from linoleic acid.

IV. Effects of Fatty Acids on the Invasive Properties of Breast Cancer Cells

A. The Metastatic Cascade

The sequence of biological events which leads to the establishment of systemic metastases is complex; only a very small minority of tumor cells are capable of completing the multistep process of tumor-host interactions. The metastatic phenotype incorporates the ability to invade into the surrounding tissue, stimulate new blood vessel formation (angiogenesis), penetrate lymphatics or blood vessels, extravasate into the host target tissue site and there proliferate to form a metastatic tumor cell colony.[30-32] A deficiency in any one of these properties will result in failure of the metastatic process.

While a number of these events may be influenced by dietary fatty acids,[33] our own work is focusing on two which are critical to the early phase of the metastatic cascade: invasion and angiogenesis.

Invasion, the first stage of the metastatic process, involves penetration of host tissue and the extracellular matrix. Three distinct, but interlocking steps occur: attachment of the tumor cell to basement membrane, enzymatic proteinolysis, and migration of the tumor cell into the digested zone and attachment to the newly exposed matrix (Figure 4).[30]

Angiogenesis is essential for the expansion of the primary tumor mass beyond a few millimeters diameter.[34] The formation of this new capillary network both sustains tumor cell growth and provides the potentially metastatic cells access to the circulatory system. It is mediated by angiogenic factors which are released by the tumor itself and by surrounding macrophages.[34]

B. Fatty Acids and Invasion *In Vitro*

We have used a membrane invasion culture system (MICS) developed by Hendrix *et al.*[35] to examine the effects of dietary fatty acids on the invasive properties of MDA-MB-435 human breast cancer cells *in vitro*.[36] We found that linoleic acid, at concentrations of 0.25 and 0.5 µg/ml, enhanced the capacity of the cells to invade through reconstituted basement membrane, an effect which was completely blocked by the addition of 20 µg/ml of indomethacin. In contrast to the omega-6 fatty acid, EPA and DHA inhibited MDA-MB-435 cell invasion *in vitro*, when added to the assay system at concentrations of 0.25 or 0.5 µg/ml.

Local tumor cell invasion requires the secretion of proteinolytic enzymes which degrade the connective tissue extracellular matrix and components of the basement membrane. One

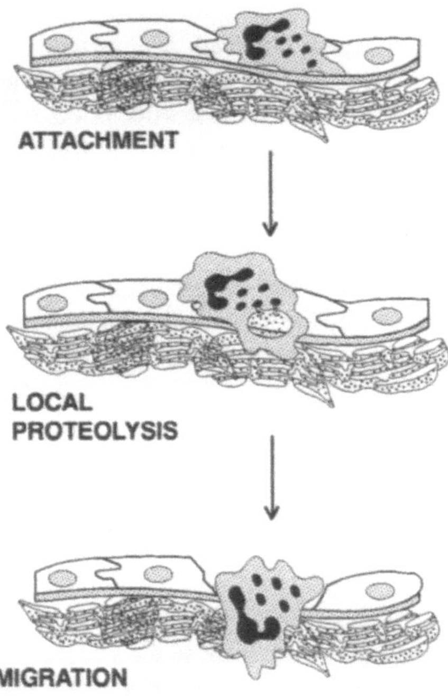

Figure 4. The three steps of tumor cell invasion.

of these enzymes, type IV collagenase, appears to play a major role in this process; for example, the capacity of rat mammary carcinoma cell sublines to metastasize correlates with their level of type IV collagenolytic activity.[37] There are, in fact, several isoenzymes with type IV collagenolytic activity,[38] but two with molecular weights of 72 and 92 kDa have received particular attention in the context of invasion and metastasis.[39-41] The 92kDa isoform, in particular, has been associated with a propensity for metastasis,[42] and with undifferentiated, aggressive breast cancers.[41]

We have shown that the addition of linoleic acid to serum-free medium increases both MDA-MB-435 cellular mRNA for the 92 kDa type IV collagenase and secretion of the active 92 kDa enzyme. Moreover, as in the case of the linoleic acid-enhanced capacity for invasion, this stimulation is blocked by indomethacin.[43]

V.　Commentary

The studies reviewed here provide support for the concept that dietary intervention based on a reduction in total fat intake, and a modification in the proportions of the different types of fat consumed, has a place in the treatment of the postsurgical breast cancer patient. However, as the results of mechanistic studies accumulate, they are suggesting some complementary approaches which, on present evidence, might include dietary supplementation with one or more omega-3 fatty acid, and treatment with selective inhibitors of eicosanoid synthesis from omega-6 fatty acid. Although indomethacin is used for the treatment of severe arthritic diseases, it can produce toxic side-effects, and it also lacks specificity with respect to its inhibitory effects on eicosanoid biosynthesis. More selective inhibitors are available, however, and are undergoing evaluation as chemopreventive agents. In this context, it may turn out that the desired target is the lipoxygenase; we have already noted that selective inhibitors of 5-lipoxygenase block linoleic acid-stimulated breast cancer cell growth,[11] and lipoxygenase products regulate tumor cell adhesion to endothelial cells and the subendothelial matrix.[44,45]

Acknowledgments

We thank Arlene Banow for assistance in preparing this manuscript. Our own work is supported by PHS Grant No. CA53124 from the National Cancer Institute, and Grant No. CN-100 from the American Cancer Society, and Grant 94A12 from the American Institute for Cancer Research.

References

1. Wynder, E.L. and L.A. Cohen, A rationale for dietary intervention in the treatment of post-menopausal breast cancer patients, *Nutr. Cancer.* 3:195 (1982).
2. Wynder, E.L., A. Morabia, D.P. Rose, and L.A. Cohen, Clinical trials of dietary interventions to enhance cancer survival, *in: Recent Progress in Research on Nutrition and Cancer*, C.J. Metlin and K. Aoki, eds., Wiley-Liss, New York, (1990).
3. Holm, L.-E., E. Nordevang, M.-L. Hjalmar, E. Lidbrink, E. Callmer, and B. Nilsson, Treatment failure and dietary habits in women with breast cancer, *J. Natl. Cancer Inst.* 85:32 (1993).
4. Dwyer, J.T., Dietary fat and breast cancer: testing interventions to reduce risks, *in: Exercise, Calories, Fat, and Cancer*, M.M. Jacobs, ed., Plenum Press, New York (1992).
5. Chlebowski, R.T., D.P. Rose, I.M. Buzzard, G.L. Blackburn, W. Insull Jr., M. Grosvenor, R. Elashoff, and E.L. Wynder, Adjuvant dietary fat intake reduction in postmenopausal breast cancer patient management, *Breast Cancer Res. Treat.* 20:73 (1992).
6. Cohen, L.A., D.P. Rose, and E.L. Wynder, A rationale for dietary intervention in postmenopausal breast cancer patients: an update, *Nutr. Cancer* 19:1 (1993).
7. Hubbard, N.E. and K.L. Erickson, Enhancement of metastasis from a transplantable mouse mammary tumor by linoleic acid, *Cancer Res.* 47:6171 (1987).
8. Katz, E.B. and E.S. Boylan, Stimulatory effect of a high polyunsaturated fat diet on lung metastasis from the 13762 mammary adenocarcinoma in female retired breeder rats, *J. Natl. Cancer Inst.* 79:351 (1987).
9. Hubbard, N.E., R.S. Chapkin, and K.L. Erickson, Inhibition of growth and linoleate-enhanced metastasis of a transplantable mouse mammary tumor by indomethacin, *Cancer Lett.* 43:111 (1988).
10. Rose, D.P. and J.M. Connolly, Stimulation of growth of human breast cancer cell lines in culture by linoleic acid, *Biochem. Biophys. Res. Commun.* 164:277 (1989).
11. Rose, D.P. and J.M. Connolly, Effects of fatty acids and inhibitors of eicosanoid synthesis on the growth of a human breast cancer cell line in culture, *Cancer Res.* 50:7139 (1990).
12. Buckman, D.K., N.E. Hubbard, and K.L. Erickson, Eicosanoids and linoleate-enhanced growth of mouse mammary tumor cells, *Prostaglandins Leukot. Essent. Fatty Acids* 44:177 (1991).
13. Price, J.E., A. Polyzos, R.D. Zhang, and L.M. Daniels, Tumorigenicity and metastasis of human breast carcinoma cell lines in nude mice, *Cancer Res.* 50:717 (1990).
14. Meschter, C.L., J.M. Connolly, and D.P. Rose, Influence of regional location of the inoculation site and dietary fat on the pathology of MDA-MB-435 human breast cancer cell-derived tumors grown in nude mice, *Clin. Exp. Metastasis* 10:167 (1992).
15. Welch, D.R., A. Neri, and G.L. Nicolson, Comparison of "spontaneous" and "experimental" metastasis using rat 13762 mammary adenocarcinoma metastatic cell clones, *Invasion Metastasis* 3:65 (1983).
16. Rose, D.P., J.M. Connolly, and C.L. Meschter, Effect of dietary fat on human breast cancer growth and lung metastasis in nude mice, *J. Natl. Cancer Inst.* 83:1491 (1991).

17. *Exercise, Calories, Fat, and Cancer*, M.M. Jacobs, ed., Plenum Press, New York (1992).

18. Rose, D.P., M.A. Hatala, J.M. Connolly, and J. Rayburn, Effects of diets containing different levels of linoleic acid on human breast cancer growth and lung metastasis in nude mice, *Cancer Res.* 53:4686 (1993).

19. Jurkowski, J.J. and W.T. Cave Jr., Dietary effects of menhaden oil on the growth and membrane lipid composition of rat mammary tumors, *J. Natl. Cancer Inst.* 74:1145 (1985).

20. Cave, W.T. Jr., ω3 fatty acid diet effects on tumorigenesis in experimental animals, *in: Health Effects of ω3 Polyunsaturated Fatty Acids in Seafoods*, A.P. Simopoulos, R.R. Kifer, R.E. Martin, and S.M. Barlow, eds., Karger, Basel (1991).

21. Karmali, R.A., J. Marsh, and C. Fuchs, Effect of omega-3 fatty acids on growth of a rat mammary tumor, *J. Natl. Cancer Inst.* 73:457 (1984).

22. Gabor, H. and S. Abraham, Effect of dietary menhaden oil on tumor cell loss and the accumulation of mass of a transplantable adenocarcinoma in BALB/c mice, *J. Natl. Cancer Inst.* 76:1223 (1986).

23. Kort, W.J., I.M. Weijma, A.M. Bijma, W.P. van Schalkwijk, A.J. Vergroesen, and D.L. Westbroek, Omega-3 fatty acids inhibiting the growth of a transplantable rat mammary adenocarcinoma, *J. Natl. Cancer Inst.* 79:593 (1987).

24. Pritchard, G.A., D.L. Jones, and R.E. Mansel, Lipids in breast cancer, *Br. J. Surg.* 76:1069 (1989).

25. Borgeson, C.E., L. Pardini, R.S. Pardini, and R.C. Reitz, Effects of dietary fish oil on human mammary carcinoma and on lipid-metabolizing enzymes, *Lipids* 24:290 (1989).

26. Gonzalez, M.J., R.A. Schemmel, J.I. Gray, L. Dugan Jr., L.G. Sheffield, and C.W. Welsch, Effect of dietary fat on growth of MCF-7 and MDA-MB-231 human breast carcinomas in athymic nude mice: relationship between carcinoma growth and lipid peroxidation product levels, *Carcinogenesis* 12:1231 (1991).

27. Rose, D.P. and J.M. Connolly, Effects of dietary omega-3 fatty acids on human breast cancer growth and metastasis in nude mice, *J. Natl. Cancer Inst.* 85:1743 (1993).

28. Croft, K.D., L.J. Beilin, F.M. Legge, and R. Vandongen, Effects of diets enriched in eicosapentaenoic or docosahexaenoic acid on prostanoid metabolism in the rat, *Lipids* 22:647 (1987).

29. Lee, J.H., M. Sugano, and T. Ide, Effects of various combinations of ω3 and ω6 polyunsaturated fats with saturated fat on serum lipid levels and eicosanoid production in rats, *J. Nutr. Sci. Vitaminol. (Tokyo)* 34:117 (1988).

30. Stracke, M.L. and L.A. Liotta, Multi-step cascade of tumor cell metastasis, *In Vivo* 6:309 (1992).

31. Radinsky, R. and I.J. Fidler, Regulation of tumor growth at organ-specific metastases, *In Vivo* 6:325 (1992).

32. Kohn, E., Aging issues in invasion and metastasis, *Cancer* 71:552 (1993).

33. Erickson, K.L. and N.E. Hubbard, Dietary fat and tumor metastasis, *Nutr. Rev.* 48:6 (1990).

34. Folkman, J., What is the evidence that tumors are angiogenesis dependent?, *J. Natl. Cancer Inst.* 82:4 (1990).

35. Hendrix, M.J.C., E.A. Seftor, R.E.B. Seftor, and I.J. Fidler, A simple quantitative assay for studying the invasive potential of high and low human metastatic variants, *Cancer Lett.* 38:137 (1987).

36. Connolly, J.M. and D.P. Rose, Effects of fatty acids on invasion through reconstituted basement membrane ('Matrigel') by a human breast cancer cell line, *Cancer Lett.* 75:137 (1993).

37. Nakajima, M., D.R. Welch, P.N. Belloni, and G.L. Nicolson, Degradation of basement membrane type IV collagen and lung subendothelial matrix by rat mammary adenocarcinoma cell clones of differing metastatic potentials, *Cancer Res.* 47:4869 (1987).

38. Stetler-Stevenson, W.G., Type IV collagenases in tumor invasion and metastasis, *Cancer Metastasis Rev.* 9:289 (1990).

39. Kossakowska, A.E., S.J. Urbanski, A. Watson, L.J. Hayden, and D.R. Edwards, Patterns of expression of metalloproteinases and their inhibitors in human malignant lymphomas, *Oncol. Res.* 5:19 (1993).

40. Seftor, R.E.B., E.A. Seftor, W.G. Stetler-Stevenson, and M.J.C. Hendrix, The 72 kDa type IV collagenase is modulated via differential expression of $\alpha v \beta 3$ and $\alpha 5 \beta 1$ integrins during human melanoma cell invasion, *Cancer Res.* 53:3411 (1993).

41. Davies, B., D.W. Miles, L.C. Happerfield, M.S. Naylor, L.G. Bobrow, R.D. Rubens, and F.R. Balkwill, Activity of type IV collagenases in benign and malignant breast disease, *Br. J. Cancer* 67:1126 (1993).

42. Yamagata, S., Y. Ito, R. Tanaka, and S. Shimizu, Gelatinases of metastatic cell lines of murine colonic carcinoma as detected by substrate gel electrophoresis, *Biochem. Biophys. Res. Commun.* 151:158 (1988).

43. Liu, X.-H. and D.P. Rose, Stimulation of type IV collagenase expression by linoleic acid in a metastatic human breast cancer cell line, *Cancer Lett.* 76:71 (1994).

44. Grossi, I.M., L.A. Fitzgerald, L.A. Umbarger, K.K. Nelson, C.A. Diglio, J.D. Taylor, and K.V. Honn, Bidirectional control of membrane expression and/or activation of the tumor cell IRGpIIb/IIIa receptor and tumor cell adhesion by lipoxygenase products of arachidonic acid and linoleic acid, *Cancer Res.* 49:1029 (1989).

45. Tang, D.G., I.M. Grossi, Y.Q. Chen, C.A. Diglio, and K.V. Honn, 12(S)-HETE promotes tumor-cell adhesion by increasing surface expression of $\alpha v \beta 3$ integrins on endothelial cells, *Int. J. Cancer* 54:102 (1993).

Chapter 9

Meta-Analysis of Animal Experiments: Elucidating Relationships Between Dietary Fat and Mammary Tumor Development in Rodents

LAURENCE S. FREEDMAN and CAROLYN K. CLIFFORD

I. Introduction

Meta-analysis is a statistical method of quantitatively combining results from different studies pertaining to a specific research question.[1] The method has also been called 'overview,' 'quantitative review,' and 'pooling.' There is now a large and rapidly growing literature on its use in medical research. It has been particularly successfully used in summarizing the results of groups of randomized clinical trials that have been designed to address the same or almost the same therapeutic question. Two pioneering examples are the meta-analyses of trials assessing the effect of beta-blockers on mortality following myocardial infarction,[2] and of tamoxifen on breast cancer recurrence and survival rates following surgery for early stage breast cancer.[3] Use of meta-analysis in epidemiology is also increasing; an interesting example is the meta-analysis of case-control studies of dietary factors and breast cancer reported by Howe *et al.*[4] which showed a significant relationship between dietary fat intake and postmenopausal breast cancer.

In this paper, we describe and illustrate how meta-analysis of data from animal experiments can elucidate relationships pertaining to dietary fat intake and mammary tumor development in rodents. Our use of meta-analysis is motivated by similar concerns to those underlying its use in clinical trials, but with a slightly different emphasis. In clinical trials, the main motivation for meta-analysis is the gain in statistical power for studying the treatment effect, obtained by the increased number of patients studied in the combined analysis. While this is also an advantage in the meta-analysis of animal experiments, other motivations lie in the ability of the combined analysis to describe broad trends of results that apply over a range of experimental conditions, to reconcile different interpretations of the

---

Laurence S. Freedman and Carolyn K. Clifford ● Biometry and Diet and Cancer Branches, Division of Cancer Prevention and Control, National Cancer Institute, Bethesda, Maryland

Diet and Breast Cancer, Edited under the auspices of the American Institute for Cancer Research, Plenum Press, New York, 1994

range of experimental results reported, and also, as explained elsewhere,[5] to take data from individual experiments that by design cannot estimate a specific relationship and combine them in an analysis which allows that relationship to be estimated.

We will describe two sets of meta-analyses, one addressing the relationships between dietary fat, energy and body weight with mammary tumor development, and the other addressing the influence of the source and type of fat intake on mammary tumors. The former analyses have been described in detail previously;[6] we repeat here the salient features. The set of analyses dealing with type and source of fat will be described in detail elsewhere; we give here the main results.

II. Data

We have built a database of results from mouse or rat experiments that have been reported in the literature up to and including the year 1987.[6] We developed the database by literature search (MEDLINE), identifying articles previous to 1966 by references in reviews. Each article identified was perused and data were included in the database if the experiment (a) included at least two groups of rodents fed diets comprising different amounts or sources of fat; (b) reported the composition of the diets; (c) reported the proportion of animals in each group developing a tumor; and (d) did not employ other interventions (e.g. immunotherapy) likely to further modify tumor incidence. Using these criteria, we extracted data from 68 articles on 114 experiments comprising a total of 11,033 animals.

In these experiments, three main types of design were employed: animals were fed *ad libitum* diets with different levels of fat, or *ad libitum* diets with different sources of fat, or were fed diets with the same source and level of fat but at different levels of energy restriction. Some experiments employ a combination of these dietary manipulations. Each meta-analysis includes a subset of experiments according to whether the experimental design fits the question being addressed. Thus some analyses include only *ad libitum* feeding experiments, some include only those experiments in which the source of fat is not varied, and so on. For each analysis, we will describe which experiments are included and why.

The main items of data that we analyzed were: the proportion of tumor-bearing animals in a group (p), the percent calories from fat in the diet (PCF), the average total energy intake per day (TCAL), the average energy from fat intake per day (FCAL), the average body weight at the end of the experiment (BW), the percent calories from linoleic acid (PLA), the percent calories from other fatty acids (POF), and the source of fat. We use the above mnemonics for these variables in our description of the analyses below.

III. Fat, Energy and Body Weight: Questions

During the 1940's and early 1950's Tannenbaum conducted extensive investigations into the relationship of dietary fat intake, energy intake and mammary tumors in mice. His experiments seemed to show that mice fed *ad libitum* diets with high levels of fat developed more tumors than those fed low levels of fat,[7] and that animals fed diets low in energy developed fewer tumors than those fed higher energy diets.[8] Later, epidemiological observations of higher breast cancer incidence rates in countries with higher per capita fat consumption[9] led to a resurgence of such experiments in rodents. However, by the end of the 1980's some disagreement between the interpretation of the results of these many experiments emerged. For example, Carroll[10] stated: "Evidence accumulated over more than 40 years has clearly established that rats and mice fed high-fat diets *ad libitum* develop tumors of the mammary gland more readily than do those fed low-fat diets", whereas a year later Boutwell[11] wrote: "Cancer incidence in specific experimental models is not dependent on the percentage of fat....Rather, the level of caloric intake versus caloric expenditure determines cancer incidence". Furthermore an analytic review by Albanes[12] was interpreted as showing that energy intake, not fat, determined mammary tumor incidence.[11]

The source of the controversy seemed to lie in three separate questions:

(i) Do animals, fed *ad libitum* diets containing different levels of fat, retain the same amount of energy?

If the answer is yes, then clearly Carroll's observations regarding *ad libitum* feeding experiments would imply that fat itself enhances mammary tumor development. If the answer is no, however, and those animals fed higher fat diets *ad libitum* retain more energy, then a second question arises:

(ii) Could the differences in energy retention explain the differences in tumor incidence that are observed in the *ad libitum* feeding experiments?

The third question arises from the contention that the effect of moderate energy restriction on mammary tumor incidence is far greater than the effect of moderate fat restriction. This question can be rephrased as:

(iii) What are the relative magnitudes of the 'energy effect' and the 'fat effect'?

These three questions can be addressed directly by a meta-analysis. One advantage of so doing is that the answers are thereby provided from the widest experience, rather than from a few selected experiments that may have been performed under special conditions.

IV. Fat, Energy and Body Weight: Answers

Each question above is addressed by specifying a statistical model and then using the data to estimate certain key quantities in the model. We will describe the model for Question (i) in full, and the models for the other questions more briefly.

(i) Do animals, fed *ad libitum* diets containing different levels of fat, retain the same amount of energy?

The model is given by:

$$\log BW_{ij} = M + EXP_i + \beta.PCF_{ij}$$

where BW_{ij} is the average body weight in the jth group of the ith experiment, M is an overall intercept, EXP_i is an experiment-specific intercept, PCF_{ij} is the percent calories from fat fed to the jth group of the ith experiment, and β is the coefficient describing how body weight depends on the level of fat in the diet. If β is zero, then body weight is not affected by the level of fat; if β is positive, then body weight increases with the level of fat.

Using data from *ad libitum* feeding experiments in which the level but not the source of fat was varied and body weight was reported (68 experiments, 159 animal groups), the estimate of β was 0.00101 with a standard error of 0.00024 (P<0.001). Thus we conclude that, on average, high-fat diets fed *ad libitum* lead to heavier body weight than do low-fat diets. However, the effect is not large. For every extra 10% calories from fat in the diet, the body weight is increased by an average of 1%, for example by 3 g in a 300 g Sprague-Dawley rat. This result leads to the second question.

(ii) Could the differences in energy retention explain the differences in tumor incidence that are observed in the *ad libitum* feeding experiments?

The model is given by:

$$\log(p_{ij}/(1-p_{ij})) = M + EXP_i + \beta_1.\log BW_{ij} + \beta_2.PCF_{ij}$$

where the term on the left hand side is the log odds of tumor incidence in the jth group of the ith experiment, β_1 is the coefficient describing the influence of body weight (and hence energy retention) on tumor incidence, and β_2 is the effect of dietary fat level on tumor incidence. If β_1 is zero then the observed differences in body weight do not explain differences in tumor incidence; a positive value for β_1 would indicate that at least some of the difference in tumor incidence between animals fed high and low-fat diets was explained by differences in energy retention. The value of β_2 indicates the effect of dietary fat level on tumor incidence, having adjusted for differences in body weight.

Using the same data as for the previous analysis the estimate of β_1 was 1.5 with a standard error of 1.2 (P=0.2) and was not significantly different from zero. In contrast the estimate of β_2 was 0.035 with a standard error of 0.003 (P<0.0001) and was clearly positive. Thus we conclude that the level of fat intake clearly influences the mammary tumor incidence and this effect is not explained in any significant manner by the observed differences in body weights of animals fed high and low-fat diets. The size of the fat coefficient β_2 indicates that on average an extra 10% calories from fat in the diet would raise tumor incidence from a baseline of 50% to a level of 59%. As shown, this effect is obtained by change in level of fat, rather than by change in energy consumption or energy retention. We now proceed to the question of how much change in energy consumption is needed to effect a similar increase (e.g. from 50% to 59%) in tumor incidence.

(iii) What are the relative magnitudes of the 'energy effect' and the 'fat effect'?

The model is given by:

$$\log(p_{ij}/(1-p_{ij})) = M + EXP_i + \beta_1.TCAL_{ij} + \beta_2.FCAL_{ij} .$$

In this model, the coefficient β_1 represents the effect upon tumor incidence of increasing non-fat intake by 1 kcal/day, and β_2 represents the effect of substituting 1 kcal of fat for 1 kcal of non-fat per day. For further discussion of the interpretation of such models see Kipnis et al.[13]

We estimated β_1 and β_2 for two sets of experiments: those in Sprague-Dawley rats fed corn oil (43 experiments, 104 animal groups) and those in mice fed any source of fat (17 experiments, 57 groups). Experiments could involve only *ad libitum* feeding or both *ad libitum* and restricted feeding comparisons. The source of fat in any experiment was required to be the same. Table 1 shows the estimates of β_1 and β_2 in these two sets of experiments. All estimates are highly statistically significant, indicating that in both sets of experiments there is strong evidence of both an 'energy effect' and a 'fat effect.' The estimated coefficients are about 5 times larger for the mice than for the Sprague-Dawley rats. This accords well with the fact that the energy intake requirement for mice is about one-fifth of that for Sprague-Dawley rats.

In each set of experiments the estimate of β_1 is approximately 1.5 times larger than the estimate of β_2. To understand the meaning of this result, consider a diet for a Sprague-Dawley rat consisting of 30 kcal of non-fat and 20 kcal of fat per day. Approximately the same reduction in tumor incidence would be achieved by substituting 10 kcal of non-fat for 10 kcal of fat per day (i.e. reducing the percent calories from fat from 40% to 20% while keeping energy intake constant) as by reducing overall energy intake by 7.7 kcal (a 15.4% reduction) while maintaining the same 40% calories from fat in the diet.

Table 1. Results of Analysis of Question(iii)

Data set	Energy		Fat	
	Estimate of β_1	z^a	Estimate of β_2	z
SD[b] rats fed corn oil	0.125	6.9	0.081	13.5
Mice	0.627	12.7	0.402	8.8

[a]   z = estimate divided by its standard error.
[b]   SD = Sprague-Dawley.

Another way of looking at these results is to note that, for Sprague-Dawley rats, an extra 5 kcal/day of non-fat would increase tumor incidence from a baseline of 50% to a level of 65%, whereas an extra 5 kcal/day of fat would increase the tumor incidence to 74%. In summary, these meta-analyses support the view that increased intake of energy from fat or non-fat sources enhances tumor development, and that the degree of enhancement is greater when the energy source is fat.

V. Type and Source of Fat: Questions

Investigators have noted apparent differences between the effects of different sources of fat on mammary tumor development.[14] Some sources appear more potent enhancers of mammary tumor development than others, while certain sources, notably fish oil,[15] appear protective. It has been reported that fats with high linoleic acid content are generally strong enhancers of mammary tumor development.[16] Meta-analysis allows a systematic investigation of this relationship. Our approach to studying the effects of source and type of fat is to pose two further questions:
  (iv) What is the quantitative relationship between the effect of the source of fat and its linoleic acid content?
This analysis will help us to find out which sources of fat appear to conform with a common relationship between tumor effect and linoleic acid, and which sources are outliers. The second part of our approach is to study whether non-linoleic fatty acids may also affect mammary tumor development. In particular we ask:
  (v) What are the relative magnitudes of the linoleic and non-linoleic fatty acid effects?
Finding a significant non-linoleic fatty acid effect might encourage a more detailed search for specific fatty acids other than linoleic that enhance mammary tumor development.

VI. Type and Source of Fat: Answers

As above, each of the questions is addressed using a statistical model.
  (iv) What is the quantitative relationship between the effect of the source of fat and its linoleic acid content?
The model is given by:
$$\log(p_{ij}/(1-p_{ij})) = M + EXP_i + \beta_k.PCF_{ij}$$
where β_k is the effect of the source of fat (source k) that is fed (at level PCF_{ij}) to the jth group of the ith experiment. We were able to estimate the effect of eleven sources of fat from our database, as reported below.

The data used for this analysis were from experiments employing *ad libitum* feeding of rats where the fat in the diet was from a single source (69 experiments, 184 groups). Mixtures of fat sources were not included in this analysis. Most of the animal groups in this analysis with a single source of fat of low linoleic acid content were thus fed a diet that was deficient in essential fatty acids. The analysis of Question (v) below addresses this problem.

Figure 1 displays the estimate of β_k for eleven sources of fat plotted against the linoleic acid content. A statistically significant relationship is seen. However, there are two obvious outliers, fish oil and primrose oil. The estimated coefficient for fish oil is significantly negative, indicating a protective effect, while primrose oil has a positive (non-significant) coefficient that is much lower than would be expected from its linoleic acid content.

This analysis suggests that linoleic acid content is a major determinant of the effect of a fat source on mammary tumor development. We next inquire whether, among fats other than fish or primrose oil, there is any effect of non-linoleic fatty acids, and we estimate the magnitude of this effect relative to that of linoleic acid.

(v) What are the relative magnitudes of the linoleic acid and non-linoleic fatty acid effects?

The model is given by:

$$\log(p_{ij}/(1-p_{ij})) = M + EXP_i + \beta_1.PLA_{ij} + \beta_2.POF_{ij}$$

where PLA_{ij} is percent calories from linoleic acid in the diet fed to the jth group of the ith experiment, POF_{ij} is the corresponding percent calories from non-linoleic fatty acids, and β_1 and β_2 are the effects on mammary tumor incidence of linoleic acid and non-linoleic fatty acids respectively.

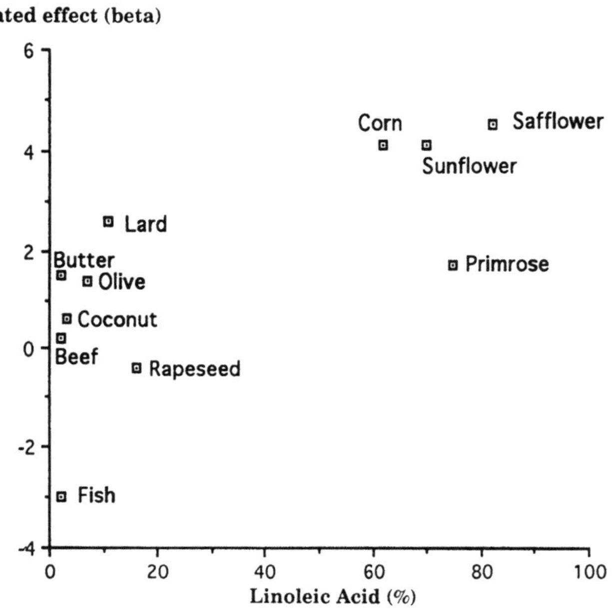

Figure 1. Relation between the estimated mammary tumor-enhancing effect of eleven sources of fat and their linoleic acid content.

The data for this meta-analysis were drawn from the same experiments as the analysis of Question (iv) but with the addition of experiments employing diets which included mixed sources of fat (76 experiments, 221 groups). As noted above, certain of the groups included in the analysis of Questions (iv) and (v) were fed diets deficient in essential fatty acids. To check on the effect of including these groups we repeated the analysis of Question (v) including only groups fed diets containing more than 2% by weight linoleic acid (71 experiments, 148 groups).

The estimates of the linoleic acid and non-linoleic fatty acid effects, β_1 and β_2, are shown in Table 2. Including groups fed diets containing less than 2% linoleic acid, the estimated linoleic acid effect was highly significant and approximately 6 times greater than the non-linoleic fatty acid effect. The estimated non-linoleic effect was also statistically significant. These results remained qualitatively similar when groups fed diets containing less than 2% linoleic acid were excluded from the analysis, although the estimate for the

Table 2. Results of Analysis of Question(v)

| Data set | Linoleic acid | | Non-linoleic acid | |
	Estimate of β_1	z^a	Estimate of β_2	z
Including <2% EFA[b]	0.061	17.3	0.009	2.6
Excluding <2% EFA	0.054	11.4	0.015	2.2

[a] z = estimate divided by its standard error.
[b]   EFA = essential fatty acid.

linoleic acid effect was somewhat smaller and the non-linoleic fatty acid effect somewhat larger.

These analyses reinforce the previous result suggesting that linoleic acid is a major determinant of mammary tumor development, and they reassure us that the previous result is not heavily influenced by inclusion of animal groups fed diets that were deficient in essential fatty acids. The significant effect of non-linoleic acid suggests that fatty acids other than linoleic acid enhance mammary tumor development. It is not clear whether this is a relatively low-level enhancing effect that is common to most of the commonly occurring non-linoleic fatty acids or whether it is a higher-level effect caused by one or more fatty acids that are present in small quantities in the common fat sources. Further experiments with mixtures of different sources of fat may be revealing.

VII. Future Work

Research on diet and breast cancer is a rapidly expanding field. Since our construction of the database in 1988, there have been many further publications of experiments which can be added. These extra data will provide added precision to the types of analysis already described and will also help us evaluate other questions of interest.

For example, there has been recent interest in whether the fat effect is modified by the level of caloric intake.[17-19] This question lends itself readily to meta-analysis. With regard to the type of fat, it is of general interest to estimate the dose response curve for linoleic acid to answer the question of whether a saturation effect can be seen beyond a certain level of intake, as has been previously suggested.[20] It will also be possible to examine the suggestion that the non-linoleic fatty acid effect is modified by the level of linoleic acid intake.[21]

Thus, carefully performed meta-analysis seems to add to our ability to summarize and draw inferences from laboratory research.

References

1. Light, R.J. and D.B. Pillemer, *Summing Up: The Science of Reviewing Research*, Harvard University Press, Cambridge (1984).
2. Yusuf, S., R. Peto, J. Lewis, R. Collins, and T. Sleight, Beta blockade during and after myocardial infarction: an overview of the randomized trials, *Prog. Cardiovasc. Dis.* 27:335 (1985).

3. Early Breast Cancer Trialists' Collaborative Group, Effects of adjuvant tamoxifen and of cytotoxic therapy on mortality in early breast cancer. An overview of 61 randomized trials among 28,896 women, *N. Engl. J. Med.* 319:1681 (1988).

4. Howe, G.R., T. Hirohata, T.G. Hislop, J.M. Iscovich, J.-M. Yuan, K. Katsouyanni, F. Lubin, E. Marubini, B. Modan, T. Rohan, P. Toniolo, and Y. Shunzhang, Dietary factors and risk of breast cancer: combined analysis of 12 case-control studies, *J. Natl. Cancer Inst.* 82:561 (1990).

5. Freedman, L.S., Meta-analysis of animal experiments on dietary fat intake and mammary tumors, *Stat. Med.* (in press).

6. Freedman, L.S., C.K. Clifford, and M. Messina, Analysis of dietary fat, calories, body weight and the development of mammary tumors in rats and mice: a review, *Cancer Res.* 50:5710 (1990).

7. Tannenbaum, A., The genesis and growth of tumors. III. Effects of a high-fat diet, *Cancer Res.* 2:468 (1942).

8. Tannenbaum, A., The dependence of tumor formation on the degree of caloric restriction, *Cancer Res.* 5:609 (1945).

9. Wynder, E.L., Identification of women at high risk from breast cancer, *Cancer* 24:1235 (1969).

10. Carroll, K.K., Introduction to Session I, Part B, Section 2, *Am. J. Clin. Nutr.* 45:216 (1987).

11. Boutwell, R.K., An overview of the role of diet and nutrition in carcinogenesis, *in: Nutrition, Growth and Cancer*, G.P. Trufiates and K.N. Prasad, eds., Alan R. Liss, New York (1988).

12. Albanes, D., Total calories, body weight and tumor incidence in mice, *Cancer Res.* 47:1987 (1987).

13. Kipnis, V., L.S. Freedman, C.C. Brown, A. Hartman, A. Schatzkin, and S. Wacholder, The interpretation of energy adjustment models for nutritional epidemiology, *Am. J. Epidemiol.* 137:1376 (1993).

14. Carroll, K.K., Summation: which fat/how much fat—animals?, *Prev. Med.* 16:510 (1987).

15. Jurkowski, J.J. and W.T. Cave Jr., Dietary effects of menhaden oil on the growth and membrane lipid composition of rat mammary tumors, *J. Natl. Cancer Inst.* 74:1145 (1985).

16. Ip, C., Fat and essential fatty acid in mammary carcinogenesis, *Am. J. Clin. Nutr.* 45:218 (1987).

17. Welsch, C.W., J.L. House, B.L. Herr, S.J. Eliasberg, and M.A. Welsch, Enhancement of mammary carcinogenesis by high levels of dietary fat: a phenomenon dependent on *ad libitum* feeding, *J. Natl. Cancer. Inst.* 82:1615 (1990).

18. Freedman, L.S. and C.K. Clifford, Enhancement of mammary carcinogenesis by high levels of dietary fat and its association with *ad libitum* feeding, *J. Natl. Cancer Inst.* 83:299 (1991).

19. Simon, R., Evaluating the differential effect of diet on rat carcinogenesis, *J. Natl. Cancer Inst.* 83:1261 (1991).

20. Ip, C., C.M. Carter, and M.M. Ip, Requirement of essential fatty acid for mammary tumorigenesis in the rat, *Cancer Res.* 45:1997 (1985).

21. Hopkins, G.J., T.G. Kennedy, and K.K. Carroll, Polyunsaturated fatty acids as promoters of mammary carcinogenesis in Sprague-Dawley rats by 7,12-dimethylbenz[a]anthracene-induced breast cancer in rats, *J. Nutr.* 114:1213 (1984).

Chapter 10
Vitamin A, Retinoids and Breast Cancer

RICHARD C. MOON

I. Introduction

Epidemiological investigations have indicated an inverse relationship between vitamin A intake and risk for developing cancer, and over the past several years an extensive effort has been directed towards understanding the mechanism by which vitamin A and retinoids (analogs of vitamin A) inhibit carcinogenesis.[1,2]

Experimentally, the role of vitamin A in regulating epithelial cell differentiation and maintenance was first demonstrated by Mori[3] and subsequently confirmed by Wolbach and Howe[4] approximately seventy years ago. They showed that feeding animals a diet deficient in retinoids resulted in the appearance of hyperkeratinization, squamous metaplasia and gross tumors in a variety of epithelial tissues, a process very similar to that induced by certain chemical carcinogens.[5] Furthermore, it has also been reported that exposure of vitamin A deficient animals to chemical carcinogens results in an increased incidence of preneoplastic lesions and cancer. Although the responses of various epithelia to the deficient state vary significantly, the systemic administration of retinoids reverses the process and restores the epithelium to a normal functional capacity.

Since carcinogen-induced metaplasia appears similar to that resulting from vitamin A (retinoid) deficiency and retinoids are known to reverse such conditions in many epithelial tissues, attempts have been made to study the effects of exogenous retinoids on the inhibition of induction and progression of cancers at various organ sites, including the mammary gland.[6] A major emphasis of retinoid research in our laboratory has focused on the chemo-prevention of mammary carcinogenesis, and this report will therefore summarize the salient features of retinoid inhibition of breast cancer.

II. Breast Cancer Models

Several tumor models for various target organs are available to study modulation of the carcinogenic process by exogenous factors. A majority of the models are designed to satisfy several criteria: 1) cancer should only be in the tissue of interest (target organ specific); 2) should mimic growth characteristics of the human counterpart (hormone dependence,

Richard C. Moon • Specialized Cancer Center, University of Illinois at Chicago, Chicago, Illinois

Diet and Breast Cancer, Edited under the auspices of the
American Institute for Cancer Research, Plenum Press, New York, 1994

metastasis, etc); 3) inducing agent (chemical carcinogen, radiation) should cause little cr no systemic toxicity; and 4) development of cancer should be relatively rapid (approximately 6 months). Among the existing models for experimental carcinogenesis studies, some offer specific advantages over others. Several of the tumor models utilized in our laboratory and by others for chemoprevention research are presented in Table 1.

Numerous studies have indicated that mammary cancers can be selectively induced in rats by either 7,12-dimethylbenz[a]anthracene(DMBA) or N-methyl-N-nitrosourea(MNU), and both mammary tumor models have been successfully utilized for chemoprevention studies. The MNU-induced tumor model was originally described by Gullino *et al.*[7] and subsequently modified in our laboratory.[8] DMBA, administered as a single intragastric dose at a concentration of 15 mg/ml sesame oil/rat, results in a 90-100% incidence of mammary tumors within 180 days post carcinogen; a single intravenous injection of 50 mg MNU/kg body weight (pH 5.0) will also lead to 100% mammary tumor incidence in rats during the same time period. The majority of cancers induced by these two carcinogens are ovarian hormone dependent with a small percentage of tumors remaining as hormone independent. Earlier studies on chemoprevention by retinoids were conducted with the DMBA-induced cancer model; however, the MNU-induced mammary cancer model has remained a model of choice for such studies for the following reasons: 1) DMBA-induced tumors are encapsulated and do not metastasize; 2) DMBA must be metabolized to an active form; and 3) DMBA induces a high incidence of adenomas and fibroadenomas. These complications do not arise in the

Table 1. Animal Tumor Models

Target	Species	Carcinogen	Tumor Type	Advantage
Breast	S/D Rat	MNU	Adenocarcinoma	Invasive, hormone responsive
Breast	S/D Rat	DMBA	Adenocarcinoma; Fibroadenoma	Hormone responsive, initiation/promotion
Breast	C_3H Mouse	MMTV	Adenoacanthoma	"Spontaneous", hormone independent
Urinary bladder	BDF Mouse	OH-BBN	TCC	Invasive, highly aggressive
Urinary bladder	F344 Rat	OH-BBN	Papilloma; TCC	Slow growing, histologic stages
Tracheo-bronchial	Syrian Hamster	MNU	SCC	Localication of tumor
Lung	Syrian Hamster	DEN	Papilloma; SCC; Adenosquamous carcinoma	Histologic types
Colon	F344 Rat	MAM	Adenocarcinoma	Single dose
Skin	CD_1 Mouse	DMBA-TPA	Papilloma; SCC	Initiation/promotion

MNU = N-methyl-N-nitrosourea; DMBA = 7, 12-dimethylbenz[a]anthracene; OH-BBN = N-butyl-N-(4-hydroxybutyl)nitrosamine; DEN = diethylnitrosamine; MAM = methylazoxymethanol acetate; TPA = 12-O-tetradecanoylphorbol-13-acetate; TCC = Transitional cell carcinoma; SCC = Squamous cell carcinoma; MMTV = mouse mammary tumor virus; S/D = Sprague-Dawley; F344 = Fischer 344.

MNU-induced cancer model. Retinoid chemoprevention studies on "spontaneous" mammary cancer have also been conducted in mice bearing the murine mammary tumor virus (MMTV), except for a study in which C_3H mice negative for MMTV were used.[9] Other studies in mice have involved mammary tumor induction in the BDF_1 and GR strains with DMBA and the ovarian steroids, respectively.

III.   Chemoprevention of Breast Cancer by Retinoids

The initial study relative to retinoid chemoprevention of mammary carcinogenesis was reported by Schmahl *et al.*[10] They observed that retinyl palmitate afforded little protection against DMBA-induced mammary carcinogenesis. The first evidence that retinoids may be effective chemopreventive agents against breast cancer was that reported by Moon *et al.*[11], who showed that dietary supplementation with retinyl acetate resulted in a 52% decrease in DMBA-induced mammary cancer, an effect subsequently confirmed using MNU as the carcinogen. Since the original studies with retinyl acetate, many retinoids have been evaluated for efficacy against chemically induced mammary cancer, and a summary of such work has been recently reviewed by Moon *et al.*[12]

Several conclusions can be drawn from these studies: (a) retinoids increase the latency of the first tumor appearance; (b) the overall tumor incidence is reduced, however, a more dramatic reduction occurs in tumor multiplicity; (c) some retinoids are apparently species specific, e.g., retinyl acetate, which is an effective retinoid against MNU- and DMBA-induced rat mammary cancer, is ineffective against mouse mammary carcinogenesis.

Past studies have indicated that the retinoids retinyl acetate and *N*-(4-hydroxyphenyl) retinamide (4-HPR), are the most effective in reducing mammary cancer incidence and increasing the latency of induced mammary cancers.[13] The number of carcinomas is also significantly reduced by the administration of either of these retinoids. On the other hand, 13-*cis*-retinoic acid has little effect upon the appearance of MNU-induced mammary carcinomas; retinyl methyl ether is of intermediate efficacy, although the latter compound is highly effective against DMBA-induced mammary carcinogenesis. Thus, it is readily apparent that minor alterations in the basic retinoid structure can significantly alter the activity of the molecule with respect to the inhibition of chemical carcinogenesis of the mammary gland.

Toxicity induced by a retinoid is of extreme importance in long term chemoprevention studies. As an example, retinyl acetate and 4-HPR are both effective inhibitors of chemical carcinogenesis of the rat mammary gland, but the patterns of metabolism and organ distribution of the two compounds are quite different. Chronic dietary administration of high doses of retinyl acetate results in an accumulation of retinyl esters in the liver, a process frequently accompanied by significant hepatic toxicity. Dietary administration of 4-HPR, on the other hand, results in a much higher level of retinoid in the mammary gland, but with relatively little liver accumulation.[14] Moreover, 4-HPR is active both in mice and rats. Thus, on the basis of its organ distribution, it would appear that 4-HPR is preferable to retinyl acetate for use in the prevention of experimental mammary cancer.

Since, for the most part, the exact time of carcinogen exposure in the human population is unknown, it is of clinical importance to determine how long after the carcinogenic insult retinoid administration can be delayed and still maintain chemopreventive efficacy. Generally, retinoids are most effective in inhibiting mammary carcinogenesis when administered shortly after carcinogen treatment. However, McCormick and Moon[15] have shown that retinoid treatment can be delayed in the rat mammary tumor model for as long as 4 months after carcinogen administration and still retain chemopreventive effectiveness; although in groups of animals in which retinoid treatment was initiated 5 months after carcinogen administration, the ability of retinoids to inhibit carcinogenesis was significantly diminished. These findings indicate that retinoid administration can be delayed for as long as 4 months but not for 5 months, indicating a "critical" time exists in tumor development beyond which retinoids may be ineffective. Such a delay in the rat model corresponds to approximately ten years in the human being.

In an effort to simulate more closely the clinical situation, two experiments were conducted in our laboratory in which retinoid treatment was not initiated until after the surgical removal of the first palpable tumor.[16] Very little quantitative inhibition of mammary tumorigenesis was evident until approximately 2 months following tumor excision. After that time a significantly reduced rate of tumor appearance was noted in the retinoid-treated group in comparison to control animals. These studies also suggest that retinoids suppress the progression of such early lesions. Although retinoids are generally much more effective against the development of early lesions, recent studies have indicated that at least two retinoids, temaroten[17] and 4-HPR,[18] may exert chemotherapeutic as well as chemopreventive activity, in that established mammary tumors regress in animals treated with these retinoids.

The evidence supporting the chemopreventive activity of retinoids in mammary tumor models in rats is substantial, however, only a few studies describing the use of retinoid in mammary tumorigenesis in mice have been reported. In the initial study, it was found that retinyl acetate did not influence (neither inhibited nor enhanced) tumor incidence, latency, or tumor number in C_3H-A^{vy} female mice positive for the mouse mammary tumor virus (MMTV).[19] However, in C_3H mice negative for the MMTV, it was found that the number of hyperplastic alveolar nodules (putative precancerous lesion) developing in animals receiving dietary 4-HPR was significantly less than that of control mice, while retinyl acetate did not affect noduligenesis in these experiments.[9] On the other hand, Welsch *et al.*[20] have reported an enhancement of tumor development in nulliparous and multiparous mice of the GR strain fed a diet supplemented with retinyl acetate, but inhibition with 4-HPR in C_3H mice.[21]

IV. Combination Chemoprevention of Breast Cancer

Although retinoids, as well as other agents, can significantly inhibit experimental carcinogenesis of various organs, a totally effective chemopreventive agent that can reduce the cancer incidence to zero is yet to be developed. It is, however, possible to enhance the inhibition achieved by retinoids by treatment with a combination of retinoid and other modulators of carcinogenesis, as demonstrated by several investigators. In most cases, combined treatment affords greater protection against cancer development than either treatment alone. Carcinogen-induced rat mammary cancer models are subject to inhibition by both retinoids and modification of host hormonal status, and it is well established that ovarian hormone-dependent tumors regress following ovariectomy of the tumor-bearing animal. Similarly, if animals are ovariectomized shortly after carcinogen administration, only the ovarian hormone-independent tumors appear, and cancer incidence is low.

The combination of ovariectomy and retinyl acetate results in a synergistic inhibition of tumor incidence and multiplicity.[22] Similar results are obtained with 4-HPR. In a more recent study, it has been shown that tamoxifen and 4-HPR, when used in combination, are much more effective in inhibiting mammary carcinogenesis than either agent alone.[23] A similar effect has recently been demonstrated for a combination of 4-HPR and dehydroepian-drosterone.[24]

The synergistic inhibition of MNU-induced mammary carcinogenesis by the concomitant administration of retinyl acetate and 2-bromo-α-ergocryptine, an inhibitor of pituitary prolactin secretion, has also been demonstrated.[25] Since the blood prolactin levels of the retinyl acetate-treated rats were similar to those of control animals, the enhanced combination effect was probably not due to a further suppression of prolactin secretion but to an effect at the level of the mammary parenchymal cell. Although hormonal modification of experimental mammary tumorigenesis is well established, this evidence indicates that the retinoids also effectively alter mammary tumorigenesis. These data suggest the existence of populations of preneoplastic and/or neoplastic cells displaying differential sensitivity to the retinoids and hormones. Whether retinoids preferentially suppress the growth of hormone-independent cell populations, reverse the neoplastic potential of these cells, or induce terminal differentiation of preneoplastic cells, as has been demonstrated in C_3H 10T1/2 cells,[26] is presently unknown.

Combination chemoprevention has also been demonstrated with retinoids and other agents that inhibit development of mammary cancer. Thompson *et al.*[27] were the first to show an enhanced inhibition of MNU-induced rat mammary carcinogenesis with retinyl acetate and selenium. The effect was confirmed by Ip and Ip[28] using the DMBA-induced mammary tumor model. Although both groups of workers found that the combined effect of retinyl acetate and selenium was substantially greater than the effect of either treatment alone, both studies were complicated by the significant reduction in food intake and body weight gain in animals receiving these chemopreventive agents. Attempts to combine modalities for prevention of mammary cancer are not always successful. For example, HPR and MVE-2 (a maleic anhydride-divinyl ether copolymer), each an immunostimulatory agent, are both effective inhibitors of MNU-induced mammary carcinogenesis in rats. However, the combination of HPR and MVE-2 was no more effective in inhibiting cancer than was either agent alone.[29]

V. Clinical Studies

Although several retinoids have been identified which posses chemopreventive activity in animal tumor models, only a few are being evaluated as chemopreventive agents for reducing cancer incidence in humans.[30] Such intervention trials are in the early stages and data relative to the effectiveness of these agents are presently not at hand. For the most part, the ongoing clinical intervention trials are "site directed" in that the end point of each trial is primarily a decreased risk of a particular cancer (e.g., lung cancer). Most of these studies involve the natural retinoids, retinol, all-*trans*- and 13-*cis*-retinoic acid, and β-carotene, which is metabolized to retinol. Some are also combination studies using vitamins C and E in addition to vitamin A (β-carotene). Only one synthetic retinoid is being evaluated clinically and that is *N*-(4-hydroxyphenyl)retinamide. This study is being conducted in Milan and is evaluating the effectiveness of the retinoid in preventing contralateral breast cancer in Stage I patients. To date, approximately 3,000 patients have been entered into the study and compliance and drug tolerance have been exceptional (A. Costa, 1993; personal communication).

Although several trials are in progress to determine the efficacy of retinoids in preventing cancer in high risk populations, it will be 3-5 years before any definitive data are in hand. In the meantime, efforts should continue to identify promising agents for clinical trials, which in the future may prove to be true chemopreventive agents applicable to the general or, at least, a large segment of the population.

VI. Summary

Both epidemiologic and experimental evidence are highly suggestive of an inverse relationship between vitamin A status and cancer induction. However, protection by vitamin

A and retinoids has not been a universal finding, nor are the compounds equally effective in inhibiting cancer induction at all organ sites. Although the experimental evidence for a protective effect of retinoids is especially strong for mammary cancer,[12] the epidemiologic evidence is less convincing.[31] Chronic pharmacologic administration of retinoids may be limited by their potential toxicity; such potential toxicity is particularly important in light of the expectation that, in order to achieve effective anticarcinogenesis in a clinical setting, administration of retinoids may be required at relatively high doses for extended periods. Clinical trials to determine the efficacy of several retinoids as cancer preventive agents are only in the early stages. Meanwhile, experimentation to develop synthetic vitamin A analogs with increased activity, increased target organ specificity, and reduced toxicity must continue.

The question of increasing anticarcinogenic activity by increasing vitamin A intake in populations whose vitamin A status is normal is more difficult to assess. The vast majority of epidemiologic studies indicate that groups with a relatively high intake of carotenoids are at a reduced risk for cancer in several target tissues. However, in populations whose vitamin A status is not deficient, increases in vitamin A and carotene intake are not well correlated with increases in serum vitamin A. Thus it is not at all clear that dietary supplementation with foods rich in preformed vitamin A and/or carotenes will, in fact, elevate serum retinol concentration to the "preventive" levels or whether such an elevation in serum retinol is really of importance, particularly as it relates to breast cancer.

References

1. McCollum, E.V. and M. Davis, The necessity of certain lipids in the diet during growth, *J. Biol. Chem.* 15:167 (1913).
2. Peto, R., R. Doll, J.D. Buckley, and M.B. Sporn, Can dietary beta-carotene materially reduce human cancer rates, *Nature* 290:201 (1981).
3. Mori, S., The changes in the paraocular glands which follow the administration of diets low in fat soluble A with notes of the effects of the same diets on the salivary glands and the mucosa of the larynx and trachea, *Johns Hopkins Hosp. Bull.* 33:357 (1922).
4. Wolbach, S.D. and P.R. Howe, Tissue changes following deprivation of fat soluble vitamin A, *J. Exp. Med.* 42:753 (1925).
5. Harris, C.C., M.B. Sporn, D.G. Kaufman, J.M. Smith, F.E. Jackson, and U. Saffiotti, Histogenesis of squamous metaplasias in the hamster tracheal epithelium caused by vitamin A deficiency or benzo[a]pyrene-ferric oxide, *J. Natl. Cancer Inst.* 48:743 (1972).
6. Moon, R.C. and L.M. Itri, Retinoids and cancer, *in: The Retinoids*, M.B. Sporn, D.S. Goodman, and A.B. Roberts, eds., Raven Press, New York (1984).
7. Gullino, P.M., H.M. Pelligrew, and F.H. Grantham, N-Nitrosomethylurea as mammary gland carcinogen in rats, *J. Natl. Cancer Inst.* 54:401 (1975).
8. McCormick, D.L., C.B. Adamowski, A. Fiks, and R.C. Moon, Lifetime dose-response relationship for mammary tumor induction by a single administration of N-methyl-N nitrosourea, *Cancer Res.* 41:1690 (1981).
9. Moon, R.C., D.L. McCormick, and R.G. Mehta, Inhibition of carcinogenesis by retinoids, *Cancer Res.* (Suppl) 43:2469s (1983).
10. Schmahl, D. and M. Habs, Experiments on the influence of an aromatic retinoid on the chemical carcinogenesis in rats by butyl-butanol-nitrosamine and 1,2-dimethylhydrazine, *Arzneimittelforschung* 28:49 (1978).
11. Moon, R.C., C.J. Grubbs, and M.B. Sporn, Inhibition of 7,12-dimethylbenz[a]anthracene-induced mammary carcinogenesis by retinyl acetate, *Cancer Res.* 36:2626 (1976).
12. Moon, R.C., R.G. Mehta, and K.V.N. Rao, Retinoids and cancer in experimental animals, *in: The Retinoids: Biology, Chemistry, and Medicine*, 2nd edition, M.B. Sporn, A.B. Roberts, and D.S. Goodman, eds., Raven Press, New York (1994).
13. Moon, R.C., R.G. Mehta, and D.L. McCormick, Modulation of mammary carcinogenesis by retinoids, *in: Vitamins, Nutrition and Cancer*, K.N. Prasad, ed., Karger, Basel (1984).

14. Moon, R.C., H.J. Thompson, P.J. Becci, C.J. Grubbs, R.J. Gander, D.L. Newton, J.M. Smith, S.R. Phillips, W.R. Henderson, L.T. Mullen, C.C. Brown, and M.B. Sporn, N-(4-hydroxyphenyl)retinamide, a new retinoid for prevention of breast cancer in the rat, *Cancer Res.* 39:1339 (1979).

15. McCormick, D.L. and R.C. Moon, Influence of delayed administration of retinyl acetate on mammary carcinogenesis, *Cancer Res.* 42:2639 (1982).

16. Moon, R.C., J.F. Pritchard, R.G. Mehta, C.T. Moides, C.F. Thomas, and N.M. Dinger, Suppression of rat mammary cancer development by N-(4-hydroxyphenyl)retinamide (4-HPR) following surgical removal of first palpable tumor, *Carcinogenesis* 10:1645 (1989).

17. Teelmann, K. and W. Bollag, Therapeutic effect of the arotenoid Ro 15-0778 on chemically induced rat mammary carcinoma, *Eur. J. Cancer Clin. Oncol.* 24:1205 (1988).

18. Dowlatshahi, K., R.G. Mehta, C.F. Thomas, N. Dinger, and R.C. Moon, Therapeutic effect of N-(4-hydroxyphenyl)retinamide on N-methyl-N-nitrosourea induced mammary cancer, *Cancer Lett.* 47:187 (1989).

19. Maiorana, A. and P. Gullino, Effect of retinyl acetate on the incidence of mammary carcinomas and hepatomas in mice, *J. Natl. Cancer Inst.* 64:655 (1980).

20. Welsch, C.W., M. Goodrich-Smith, C.C. Brown, and N. Crowe, Enhancement by retinyl acetate of hormone-induced mammary tumorigenesis in female GR/A mice, *J. Natl. Cancer Inst.* 67:935 (1981).

21. Welsch, C.W., J.V. DeHoog, and R.C. Moon, Inhibition of mammary tumorigenesis in nulliparous C_3H mice by chronic feeding of the synthetic retinoid, N-(4-hydroxyphenyl)retinamide, *Carcinogenesis* 4:1185 (1983).

22. McCormick, D.L., R.G. Mehta, C.A. Thompson, N. Dinger, J.A. Caldwell, and R.C. Moon, Enhanced inhibition of mammary carcinogenesis by combination N-(4-hydroxyphenyl)retinamide and ovariectomy, *Cancer Res.* 42:509 (1982).

23. Ratko, T.A., C.J. Detrisac, N.M. Dinger, C.T. Thomas, G.J. Kelloff, and R.C. Moon, Chemopreventive efficacy of combined retinoid and tamoxifen treatment following surgical excision of a primary mammary cancer in female rats, *Cancer Res.* 49:4472 (1989).

24. Ratko, T.A., R.G. Mehta, C.J. Detrisac, G.J. Kelloff, and R.C. Moon, Inhibition of rat mammary gland chemical carcinogenesis by dietary dehydroepiandrosterone or a fluorinated analogue of dehydroepiandrosterone, *Cancer Res.* 51:481 (1991).

25. Welsch, C.W., C.K. Brown, M. Goodrich-Smith, J. Chiusano, and R.C. Moon, Synergistic effect of chronic prolactin suppression and retinoid treatment in the prophylaxis of N-methyl-N-nitrosourea-induced mammary tumorigenesis in female Sprague-Dawley rats, *Cancer Res.* 40:3095 (1980).

26. Bertram, J.S., Inhibition of neoplastic transformation *in vitro* by retinoids, *Cancer Surv.* 3:243 (1983).

27. Thompson, H.J., L.D. Meeker, and P.J. Becci, Effect of combined selenium and retinyl acetate treatment on mammary carcinogenesis, *Cancer Res.* 41:1413 (1981).

28. Ip, C. and M.M. Ip, Chemoprevention of mammary tumorigenesis by a combined regimen of selenium and vitamin A, *Carcinogenesis* 2:915 (1981).

29. McCormick, D.L., P.J. Becci, and R.C. Moon, Inhibition of mammary and urinary bladder carcinogenesis by a retinoid and a maleic anhydride-divinyl ether copolymer (MVE-2), *Carcinogenesis* 3:1473 (1982).

30. Malone, W.F., G.J. Kelloff, C. Boone, and D.W. Nixon, Chemoprevention and modern cancer prevention, *Prev. Med.* 18:553 (1989).

31. Hong, W.K. and L.M. Itri, Retinoids and human cancer, *in: The Retinoids: Biology, Chemistry, and Medicine*, 2nd edition, M.B. Sporn, A.B. Roberts, and D.S. Goodman, eds., Raven Press, New York (1994).

Chapter 11

Vitamin D Adequacy: A Possible Relationship to Breast Cancer

HAROLD L. NEWMARK

I. Introduction

Dietary fat has long been suspected of playing a significant role in the etiology of breast cancer.[1-3] This stems from marked international correlations of estimated per capita fat intake in whole populations, presumably during an entire lifetime, compared with breast cancer incidence and mortality rates,[1] as well as migrant studies,[4] and temporal increases in breast cancer incidence paralleling higher rates of fat intake.[5] In contrast, shorter term prospective or case-control studies have generally not found a positive correlation between (adult) fat intake and breast cancer risk.[6-12]

Animal studies have demonstrated that dietary fat acts as a promoter rather than initiator of chemically-induced mammary cancer[13-16] and colon cancer.[17] For the colon the hypothesis was suggested that increased dietary fat resulted in increased levels of free fatty acids and free unconjugated bile acids in the colon. Acting as irritants, these would increase colon epithelial cell proliferation, thus enhancing colon cancer risk. The hypothesis further suggested that this irritation-driven hyperproliferation could be reduced by increasing dietary calcium and vitamin D. Studies in animals, short-term interventions in humans, and epidemiologic studies support the utility of the concept that increased dietary calcium and vitamin D can reduce the risk of colon cancer related to high dietary fat intake.[18]

II. Experimental

Because of the similar epidemiological correlations between colon and mammary cancer, animal studies were designed to test the influence of dietary calcium and vitamin D levels on the promotional effect of high-fat diets in chemically induced mammary cancer. In the first series,[19] variation of dietary calcium and vitamin D had little effect with low-fat diets, except when the dietary calcium was very low. On high-fat diets, however, a significant

---

Harold L. Newmark ● Memorial Sloan-Kettering Cancer Center, New York, New York and Rutgers University College of Pharmacy, Piscataway, New Jersey

Diet and Breast Cancer, Edited under the auspices of the
American Institute for Cancer Research, Plenum Press, New York, 1994

increase in mammary tumorigenesis resulted from decreasing the dietary calcium and vitamin D to levels similar to those of human adults in North America.

In a second series[20,21] dietary levels of calcium, phosphate and vitamin D in high-fat diets were varied independently in studies on chemically-induced carcinogenesis in rats. The results suggested that both phosphate and vitamin D have interactive effects with dietary calcium. Dietary vitamin D had the largest individual effect, and in higher amounts it inhibited tumorigenesis in the presence of low amounts of calcium and phosphate, but it was less effective with a high-calcium and phosphate diet. These diet studies in rats are in general agreement with a mechanism postulated for the effect of high dietary fat in breast cancer promotion. According to this concept, high dietary fat produces an increased flux of fat in the breast, particularly during growth. Movement of fats in and out of cells in the breast, particularly in and around the cancer-sensitive ductal epithelial cells, is largely via the hydrolysis of circulating triglycerides to free fatty acids for transport across cell membranes. This presents a risk to calcium-dependent cell structures and very sensitive cell signaling systems, due to the strong avidity of free fatty acids for calcium.[22] Maintenance of proper cellular calcium levels depends on adequate circulating calcium and the active hormone form of vitamin D, calcitriol (1,25-dihydroxycholecalciferol or 1,25 di-OHD$_3$).[23] The blood-level of calcium is sufficiently important physiologically to be tightly controlled within a narrow range by interaction of parathyroid hormone, calcitonin, 1,25-di-OHD$_3$-controlled dietary absorption, losses due to gastrointestinal food component reactions, and losses in urine and sweat.[24] However, blood levels of 1,25-di-OHD$_3$ also facilitate cellular uptake of calcium from the blood. Low dietary vitamin D could, therefore, be expected to magnify the high-fat promoting effect on breast cancer by limiting the physiological capacity to elevate 1,25-di-OHD$_3$, resulting in reduced ability of cells to replenish calcium lost to free fatty-acid binding. Accordingly, it was noted with interest that in these animal studies, the greatest single effect was due to dietary vitamin D[20,21] although the data are limited.

Biologically active hormone-like forms of vitamin D, such as 1,25-di-OHD$_3$ and 1-α-hydroxycholecalciferol (1-α-OHD$_3$) have shown potent differentiating activity in cancer cell lines from a variety of tissues.[24] There have been similar reports of the differentiation effect of 1,25-di-OHD$_3$ on human breast cancer cell lines.[25,26] Studies by Colston *et al.* on human breast tumor and cancer cell lines indicated that over 80% of human breast tumors contain 1,25-di-OHD$_3$ receptors, and that *in vitro* 1,25-di-OHD$_3$, unlike estrogens, inhibits proliferation and promotes differentiation in several cell types. Patients with 1,25-di-OHD$_3$ receptor-positive tumors had significantly longer disease-free survival than those with receptor-negative tumors ($\chi^2 = 4$, p<0.05). In rats with mammary tumors induced by methylnitrosourea, treatment with the synthetic analogue 1-α-OHD$_3$ produced significant inhibition of tumor progression.[27] The levels of 1,25-di-OHD$_3$ occurring *in vivo* may exert an inhibiting effect on receptor-positive tumors.[27,28]

III. Discussion

Carcinogenesis is currently viewed as including initiating genotoxic events and agents, followed by a long period of promotion and/or progression associated with increased cell proliferation and other changes. Little is known of initiating agents in human breast cancer. However, several of the agents associated with breast cancer, such as estrogens and dietary fat, have demonstrated promotion/progression activity in chemically-initiated mammary cancer studies in animals. Zhang *et al.*[29] reported that high dietary fat significantly increased proliferation (about two-fold) of murine mammary epithelial cells, particularly in the epithelium of the terminal ducts.

There was a large increase in proliferation, magnified by high dietary fat, beginning at 4 weeks of age in mice. The high proliferative rates were maintained to about 10 weeks of age during the period of normal growth and maturation of the mammary glands, before declining to lower adult levels. This period of life corresponds roughly to puberty and adolescence of the human, a period when rapid cellular proliferation and growth of mammary epithelium are "hormonally driven" by gonad-supplied estrogens. The large increase in mammary epithelial cell proliferation caused by high dietary fat during the period of puberty and adolescence may be the key to explaining several observed phenonema:

A. In chemically induced carcinogenesis in rodents, the carcinogen must be administered at 30-45 days of age (mice) or 40-55 days of age (rats) for significant production of tumors, i.e., when the target mammary epithelium is most rapidly proliferating normally due to response to gonadal estrogens.
B. Epidemiological studies of dietary fat and breast cancer based on inter-country comparisons which estimate fat intake over the entire lifetime *including* puberty and adolescence, share a consistent association of high dietary fat with risk of breast cancer.
C. Epidemiological studies based on case control or prospective cohorts of adult women, often menopausal or postmenopausal, fail to show an association between dietary fat and risk of breast cancer. Perhaps these studies should include dietary fat, calcium and vitamin D during puberty and adolescence, periods of greatest dietary effect on the mammary tissue, even though the outcome is not manifest until much later in life.

Zhang *et al.*[29] also demonstrated that increased dietary calcium significantly reduced the increased mammary epithelial proliferation induced by high fat. The bioavailability of the dietary calcium was ensured by an optimal high level of dietary vitamin D in the AIN-76 Vitamin Mix.[29] After 20 weeks on a nutritional stress diet which "mimics" the human Western-style diet in terms of high fat and phosphate, low calcium and vitamin D, mammary ductal hyperproliferation and hyperplasia were found at 26 weeks of age.[30] Using the same stress diet, colonic epithelial hyperproliferation and hyperplasia were produced, and eliminated by increasing dietary calcium.[18] This suggests that increasing dietary calcium and vitamin D might prevent and/or reverse hyperproliferation and hyperplasia induced by the stress diet in the mammary epithelium.

The biochemistries of calcium and vitamin D are closely linked since bioavailability of calcium (e.g., absorption in the gastrointestinal tract and also into cells within the body) depends on adequate vitamin D. Dietary vitamin D in North America is low, generally well below the US RDA. However, dietary intake is "supplemented" with production in skin exposed to clear, bright sunlight, but sunlight varies considerably with season, latitude and degrees of haze and smog which act as ultraviolet blocking aerosols in the atmosphere. In this regard, the studies of Garland *et al.*[31] and Gorham *et al.*[32] suggest a strong inverse correlation between breast cancer and availability of solar radiation for *in vivo* skin production of vitamin D, especially as applied to the U.S. and Canada. The relationship to colon cancer was also reviewed.[33]

Lower solar radiation, particularly in urban areas where the greater part of the U.S. population lives, thus results in reduced biologically available vitamin D from this source and creates an increased dependency on dietary intake. The U.S. RDA for dietary vitamin D is 10 µg (= 400 units) from 1-24 years of age for both males and females, and 5 µg (= 200 units) above 24 years of age, except for pregnant females. However, daily dietary intakes are far lower: females' consumption averages 1.5 µg (60 units), and elderly females have a median intake of 1.35 µg (54 units).[34]

The low average intake of dietary vitamin D, coupled with the reduced capacity to convert vitamin D to 25-OHD$_3$[35] and further to 1,25-di-OHD$_3$[36] serves to explain the very poor status of active vitamin D forms in older populations.[37-40] However, this seems at least partly correctable with dietary supplementation.[41]

Studies on the etiology of osteoporosis bear a similarity to the studies of breast cancer in terms of dietary correlations. In a review of dietary studies and osteoporosis, the Subcommittee on the Tenth Edition of the RDAs emphasized the importance of intakes of recommended levels of calcium and its biochemically associated vitamin D throughout childhood to age 25 years.[34] There is little risk in intake of vitamin D at RDA levels, since risk of hypervitaminosis only occurs on prolonged intake of five times or more of the RDA (over 2000 units, = 50 mcg daily).[34]

IV. Summary

(1) Low levels of dietary calcium and vitamin D, biochemically interrelated, increase the promoting action of high dietary fat on chemically induced mammary carcinogenesis in animal studies.

(2) High dietary fat increases mammary epithelial cell proliferation, particularly the "hormonally driven" hyperproliferation during breast growth and development in young animals. Increased dietary calcium (and probably vitamin D) lessens the increase of proliferation induced by high fat. These data, although limited, suggest that the maximum effect of diet (high fat increase, as well as calcium and vitamin D modulation) on eventual breast cancer may be during puberty, and adolescence, when the mammary gland is actively growing and developing.

(3) An inverse epidemiological correlation has been developed between sunlight availability as a source of vitamin D and the risk of breast cancer in the U.S. and Canada.

(4) Current vitamin D and calcium dietary intake in the U.S. is far below the RDA in all female age groups, particularly for the elderly.

(5) Reduction of breast cancer risk, and simultaneously osteoporosis, might be achieved by increasing dietary intake of calcium and vitamin D to RDA levels. This may be particularly applicable to females during puberty and adolescence.

References

1. Armstrong, B. and R. Doll, Environmental factors and cancer incidence and mortality in different countries, with special reference to dietary practices, *Int. J. Cancer* 15:617 (1975).

2. Carroll, K.K. and H.T. Khor, Dietary fat in relation to tumorigenesis, *Prog. Biochem. Pharmacol.* 10:308 (1975).

3. Ingram, D.M., Trends in diet and breast cancer mortality in England and Wales 1908-1977, *Nutr. Cancer* 3:75 (1981).

4. Buell, P., Changing incidence of breast cancer in Japanese-American women, *J. Natl. Cancer Inst.* 51:1479 (1973).

5. Miller, A.B., Role of nutrition in the etiology of breast cancer, *Cancer* 39:2704 (1977).

6. Miller, A.B., A. Kelly, N.W. Choi, V. Matthews, R.W. Morgan, L. Munan, J.D. Burch, I. Feather, G.R. Howe, and M. Jain, A study of diet and breast cancer, *Am. J. Epidemiol.* 107:499 (1978).

7. Katsouyanni, K., D. Trichopoulos, P. Boyle, E. Xirouchaki, A. Trichopoulou, B. Lisseos, S. Vasilaros, and B. MacMahon, Diet and breast cancer: a case-control study in Greece, *Int. J. Cancer* 38:815 (1986).

8. Lubin, F., Y. Wax, and B. Modan, Role of fat, animal protein, and dietary fiber in breast cancer etiology: a case-control study, *J. Natl. Cancer Inst.* 77:605 (1986).

9. Wynder, E.L., D.P. Rose, and L.A. Cohen, Diet and breast cancer in causation and therapy, *Cancer* 58 (Suppl):1804 (1986).

10. Hirohata, T., A.M.Y. Nomura, J.H. Hankin, L.N. Kolonel, and J. Lee, An epidemiologic study on the association between diet and breast cancer, *J. Natl. Cancer Inst.* 78:595 (1987).

11. Jones, D.Y., A. Schatzkin, S.B. Green, G. Block, L.A. Brinton, R.G. Ziegler, R. Hoover, and P.R. Taylor, Dietary fat and breast cancer in the National Health and Nutrition Examination Study. I. Followup Study, *J. Natl. Cancer Inst.* 79:465 (1987).

12. Willett, W.C., M.J. Stampfer, G.A. Colditz, B.A. Rosner, C.H. Hennekens, and F.E. Speizer, Dietary fat and the risk of breast cancer, *N. Engl. J. Med.* 316:22 (1987).

13. Rogers, A.E. and S.Y. Lee, Chemically-induced mammary gland tumors in rats: modulation by dietary fat, *Prog. Clin. Biol. Res.* 222:255 (1986).

14. Carroll, K.K. and G.J. Hopkins, Dietary polyunsaturated fat *versus* saturated fat in relation to mammary carcinogenesis, *Lipids* 14:155 (1979).

15. Ip, C., C.A. Carter, and M.M. Ip, Requirement of essential fatty acid for mammary tumorigenesis in the rat, *Cancer Res.* 45:1997 (1985).

16. Ip, C., Fat and essential fatty acid in mammary carcinogenesis, *Am. J. Clin. Nutr.* 45 (Suppl):218 (1987).

17. Bull, A.W., B.K. Soulier, P.S. Wilson, M.T. Hayden, and M.D. Nigro, Promotion of azoxymethane-induced intestinal cancer by high-fat diet in rats, *Cancer Res.* 39:4956 (1979).

18. Newmark, H.L. and M. Lipkin, Calcium, vitamin D and colon cancer, *Cancer Res.* 52 (Suppl):2067s (1992).

19. Jacobson, E.A., K.A. James, H.L. Newmark, and K.K. Carroll, Effects of dietary fat, calcium, and vitamin D on growth and mammary tumorigenesis induced by 7,12-dimethylbenz[a]anthracene in female Sprague-Dawley rats, *Cancer Res.* 49:6300 (1989).

20. Carroll, K.K., E.A. Jacobson, L.A. Eckel, and H.L. Newmark, Calcium and carcinogenesis of the mammary gland, *Am. J. Clin. Nutr.* 54:206s (1991).

21. Carroll, K.K., L.A. Eckel, L.J. Fraher, J.V. Frei, and H.L. Newmark, Dietary calcium, phosphate and vitamin D in relation to mammary carcinogenesis, *in: Calcium, Vitamin D, and Prevention of Colon Cancer*, M. Lipkin, H.L. Newmark, and G. Kelloff, eds., CRC Press, Boca Raton (1991).

22. Newmark, H.L., M.H. Wargovich, and W.R. Bruce, Colon cancer and dietary fat, phosphates and calcium: a hypothesis, *J. Natl. Cancer Inst.* 72:1323 (1984).

23. Newmark, H.L., Calcium in cellular function, *Triangle* 28 (Suppl 1):9 (1989).

24. DeLuca, H.F., The vitamin D story: a collaborative effort of basic science and clinical medicine, *FASEB J.* 2:224 (1988).

25. Koga, M., J.A. Eisman, and R.L. Sutherland, Regulation of epidermal growth factor receptor levels by 1,25-dihydroxy vitamin D_3 in human breast cancer cells, *Cancer Res.* 48:2734 (1988).

26. Frappart, L., N. Falette, M.F. Lefebre, A. Bremond, J.L. Vauzelle, and S. Saez, *In vitro* study of effects of 1,25-dihydroxy vitamin D_3 on the morphology of human breast cancer cell line BT.20, *Differentiation* 40:63 (1989).

27. Colston, K.W., U. Berger, and R.C. Coombes, Possible role for vitamin D in controlling breast cancer cell proliferation, *Lancet* i:188 (1989).

28. Colston, K.W., S.K. Chander, A.G. MacKay, and R.C. Coombes, Effects of synthetic vitamin D analogues on breast cancer cell proliferation *in vivo* and *in vitro*, *Biochem. Pharmacol.* 44:693 (1992).

29. Zhang, L., R.P. Bird, and W.R. Bruce, Proliferative activity of murine mammary epithelium as affected by dietary fat and calcium, *Cancer Res.* 47:4905 (1987).

30. Khan, M., K. Yang, H. Newmark, R. Rivlin, and M. Lipkin, Mammary ductal epithelial cell hyperproliferation and hyperplasia induced by four components of a Western-style diet, *Proc. Am. Assoc. Cancer Res.* 34:554 (1993).

31. Garland, F., C.F. Garland, E.D. Gorham, and J.F. Young, Geographic variation in breast cancer mortality in the United States: a hypothesis involving solar radiation, *Prev. Med.* 19:614 (1990).

32. Gorham, E.D., C.F. Garland, and F. Garland, Acid haze, air pollution and breast and colon cancer mortality in 20 Canadian cities, *Can. J. Public Health* 80:96 (1989).

33. Garland, C.F. and F. Garland, *The Calcium Connection*, G.P. Putnam, New York (1988).

34. *Recommended Dietary Allowances*, 10th edition, National Academy Press, Washington, pp 92-98 (1989).

35. Baker, M.R., M. Peacock, and B.E. Nordin, The decline in vitamin D status with age, *Age Ageing* 9:249 (1980).

36. Armbrecht, H.J., T.V. Zenser, and B.B. Davis, Effect of age on the conversion of 25-hydroxy vitamin D_3 to 1,25-dihydroxy vitamin D_3 by the kidney of the rat, *J. Clin. Invest.* 66:1118 (1980).

37. Omdahl, J.L., P.J. Garry, L.A. Hensaker, W.C. Hunt, and J.S. Goodwin, Nutritional status in a healthy elderly population: vitamin D, *Am. J. Clin. Nutr.* 36:1125 (1982).

38. Payette, J. and K. Gray-Donald, Dietary intake and biochemical indices of nutritional status in an elderly population, with estimates of the precision of the seven-day food record, *Am. J. Clin. Nutr.* 54:478 (1991).

39. Delvin, E.E., A. Imbach, and M. Copti, Vitamin D nutritional status and related biochemical indices in an autonomous elderly population, *Am. J. Clin. Nutr.* 48:373 (1988).

40. Lore, F., G. DiCairano, and G. DiPerri, Vitamin D status in the extreme age of life, *Ann. Med. Interne. (Paris)* 137:209 (1986).

41. Himmelstein, S., T.L. Clemens, A. Rubin, and R. Lindsay, Vitamin D supplementation in elderly nursing home residents increases 25-(OH)D but not 1,25(OH)$_2$D, *Am. J. Clin. Nutr.* 52:701 (1990).

Chapter 12

Vitamin D and Breast Cancer

SYLVIA CHRISTAKOS

I. Discussion

The most active form of vitamin D-1,25-dihydroxyvitamin D_3 (1,25(OH)$_2$D$_3$), has been found to act analogously to a steroid hormone and to regulate calcium homeostasis by affecting intestinal calcium absorption and bone resorption.[1] However, in addition to its role in maintaining calcium homeostasis, 1,25(OH)$_2$D$_3$ also acts as a differentiating agent.[2] Studies in leukemic cells have indicated antiproliferative and prodifferentiating effects of 1,25(OH)$_2$D$_3$.[3-6] It has been reported that 1,25(OH)$_2$D$_3$ at physiological concentrations can reduce the proliferation of the human promyelocytic leukemia cell line HL-60 and the human monoblastic leukemia cell line U937 and can cause differentiation of these cells to mature macrophagelike cells.[3-6] These alterations were found to be preceded by a marked decline in c-*myc* mRNA and a transient induction in c-*fos* mRNA.[7,8] These studies suggested that 1,25(OH)$_2$D$_3$ may be useful in the treatment of acute myeloid leukemia. However, in clinical trials oral administration of 1,25(OH)$_2$D$_3$ resulted in the development of hypercalcemia before therapeutic concentrations for the treatment of leukemia could be attained.[4]

In recent studies attention has been given to the synthesis of vitamin D analogs that inhibit proliferation and induce differentiation of leukemia cells without causing hypercalcemia.[9-14] In 1987 Ostrem *et al.*[9] reported that increasing the side chain length at the 24 position afforded an analog which increased differentiative activity of HL-60 leukemia cells. Perlman *et al.* showed that 24, 24-dihomo-1,25(OH)$_2$D$_3$ (increase in side chain length of 2 carbons) is approximately 10 times more active than 1,25(OH)$_2$D$_3$ in causing differentiation of HL-60 cells.[10] However, this compound is at least 10 times less active in intestinal calcium transport than 1,25(OH)$_2$D$_3$ and is approximately 1,000 times less active in mobilizing skeletal calcium than 1,25(OH)$_2$D$_3$.[10] These findings suggest the possible importance of 24,24-dihomo-1,25(OH)D$_2$D$_3$ in controlling leukemia without inducing hypercalcemia. Zhou *et al.*[11] also tested a number of novel analogs of 1,25(OH)$_2$D$_3$ and found 1,25-dihydroxy-16-ene-23-yne vitamin D_3 (1,25(OH)$_2$-16-ene-23-yne D$_3$) was 4-12 fold more potent than 1,25(OH)$_2$D$_3$ in inhibiting growth and inducing differentiation of leukemic cells. It was less potent than 24,24-dihomo-1,25(OH)$_2$D$_3$ in absorption (30 times less potent than

Sylvia Christakos ● Department of Biochemistry and Molecular Biology, University of Medicine and Dentistry of New Jersey, New Jersey Medical School, Newark, New Jersey

Diet and Breast Cancer, Edited under the auspices of the
American Institute for Cancer Research, Plenum Press, New York, 1994

1,25-$(OH)_2D_3$) and was 50 times less potent than 1,25$(OH)_2D_3$ in bone calcium mobiliza-tion.[11] 1,25-Dihydroxy-16-ene-23-yne vitamin D_3 was also the first analog to prolong the survival time of leukemic mice, suggesting the therapeutic potential of this compound in the treatment of leukemia.[12] Recently we reported that 1,25,28-trihydroxyvitamin D_2 (1,25,28 $(OH)_3D_2$), similar to the other analogs,[9-13] has differentiating activity without hypercalcemic activity, for it is at least 60 times less potent than 1,25$(OH)_2D_3$ in intestinal calcium absorption.[14] However, unlike the 1,25$(OH)_2$24 homologs[9,10] and the analogs of 1,25$(OH)_2D_3$ possessing a 16-ene-23-yne structure,[11] high concentrations of 1,25,28$(OH)_3D_2$ are needed to induce differentiation.[14] Further studies are needed to determine the possible therapeutic potential of 1,25,28$(OH)_3D_2$.

In addition to leukemic cells, 1,25$(OH)_2D_3$ has been found to inhibit the growth of a number of other cancer cells, including breast cancer cells.[15-18] Specific receptors for 1,25$(OH)_2D_3$ have been identified in these cancer cells,[15-18] suggesting that 1,25$(OH)_2D_3$ may induce differentiation by affecting gene transcription, similar to its proposed mechanism for regulating calcium homeostasis.[1] The presence of vitamin D receptors (VDR) in breast carcinoma is correlated with improved prognosis in breast cancer patients, suggesting the possibility that 1,25$(OH)_2D_3$ may promote differentiation and/or inhibit growth of breast cancer cells *in vivo* via VDR mediated events.[19] Other *in vivo* studies using N-methyl-N-nitrosourea (MNU) treated rats[19,20] or athymic mice implanted with breast tumor tissue[21] have indicated that 1,25$(OH)_2D_3$ or analogs of 1,25$(OH)_2D_3$ (EB1089, characterized by a modification in the C17 side chain of the vitamin D molecule and 22-oxa-1,25-dihydroxy-vitamin D_3, an analog with an oxygen atom in the side chain skeleton) can inhibit the progression of mammary tumor growth. This suggests the therapeutic potential of analogs of 1,25$(OH)_2D_3$ in the treatment of breast cancer. ED1089 was found to be a more potent inhibitor of MCF-7 cell proliferation than 1,25$(OH)_2D_3$ (50% inhibition of ^{3}H-thymidine for EB1089 was 5×10^{-10}M compared to 50% inhibition by 1,25$(OH)_2D_3$ at 5×10^{-9}M).[20] When rats bearing at least one MNU-induced tumor (> 10 mm diameter) were treated orally with 0.5 µg/kg body weight (BW) EB1089/day for 28 days, mean tumor volume was approximately half that in the control group and hypercalcemia was not observed. A higher dose of EB1089 (1 µg/kg/day) resulted in more marked inhibition of tumor volume compara-ble to effects seen with tamoxifen (1 mg/kg/day). However, serum calcium significantly increased in these rats.[20] Another analog of 1,25$(OH)_2D_3$, 22-oxa-1,25-dihydroxyvitamin D_3 (OCT) inhibited the proliferation of both estrogen receptor (ER) positive and estrogen receptor negative breast cancer cells.[21] In athymic mice implanted with the ER negative MX-1 tumor, intratumor injection of OCT 3 times per week at a concentration of 1 µg/kg BW for 26 days resulted in a 70% reduction in tumor weight. The antitumor effect of 1 µg/kg OCT was found to be greater than that of 500 µg/kg BW adriamycin, while the effects of OCT and adriamycin were additive. After 26 days of combined treatment the tumor weight in treated animals was 21.7% of the vehicle treated group. This dose of OCT (1µg/kg BW) caused a marked suppression of the size and weight of the breast cancer, but it did not affect serum calcium. Since OCT was effective in treating estrogen receptor negative breast cancer, these findings suggest that OCT has potential as a new compound for the endocrine treatment of breast cancer.

II. Summary

1. 1,25-Dihydroxyvitamin D_3, the biologically active form of vitamin D, in addition to regulating calcium homeostasis, also has antiproliferative and prodifferentiating effects.
2. Most studies concerning the therapeutic potential of analogs of 1,25$(OH)_2D_3$, which are antiproliferative and prodifferentiating but do not cause hypercalcemia, have been done using leukemic cells.

3. Recent evidence from both *in vivo* and *in vitro* studies has indicated that $1,25(OH)_2D_3$ or analogs of $1,25(OH)_2D_3$ can inhibit the growth of breast cancer cells, thus suggesting the therapeutic potential of analogs of $1,25(OH)_2D_3$ in the treatment of breast cancer.

References

1. Reichel, H., H.P. Koeffler, and A.W. Norman, The role of the vitamin D endocrine system in health and disease, *N. Engl. J. Med.* 320:980 (1989).
2. Abe, E., C. Miyaura, H. Sakagami, M. Takeda, K. Konno, T. Yamazaki, S. Yoshiki, and T. Suda, Differentiation of mouse myeloid leukemia cells induced by $1\alpha,25$-dihydroxyvitamin D_3, *Proc. Natl. Acad. Sci. USA* 78:4990 (1981).
3. Munker, R., A.W. Norman, and H.P. Koeffler, Vitamin D compounds: effect on clonal proliferation and differentiation of human myeloid cells, *J. Clin. Invest.* 78:424 (1986).
4. Koeffler, H.P., K. Hirji, and L. Itri, 1,25-Dihydroxyvitamin D_3: *in vivo* and *in vitro* effect on human preleukemic and leukemic cells, *Cancer Treat. Rep.* 69:1399 (1985).
5. Dodd, R.C., M.S. Cohen, S.L. Newman, and T.K. Gray, Vitamin D metabolites change the phenotype of monoblastic U937 cells, *Proc. Natl. Acad. Sci. USA* 80:7538 (1983).
6. Tanaka, H., E. Abe, C. Miyaura, Y. Shiina, and T. Suda, $1\alpha,25$-Dihydroxyvitamin D_3 induces differentiation of human promyelocytic leukemia cells (HL-60) into monocyte-macrophages but not into granulocytes, *Biochem. Biophys. Res. Commun.* 117:86 (1983).
7. Simpson, R.U., T. Hsu, D.A. Begley, B.S. Mitchell, and B.N. Alizadeh, Transcriptional regulation of the c-myc protooncogene by 1,25-dihydroxyvitamin D_3 in HL-60 promyelocytic leukemia cells, *J. Biol. Chem.* 262:4104 (1987).
8. Brevli, Z.S. and G.P. Studzinksi, Changes in the expression of oncogenes encoding nuclear phosphoproteins but not c-Ha-*ras* have a relationship to monocytic differentiation of HL60 cells, *J. Cell. Biol.* 102:2234 (1986).
9. Ostrem, V.K., Y. Tanaka, J. Prahl, H.F. DeLuca, and N. Ikekawn, 24- and 26-homo-1,25-dihydroxyvitamin D_3: preferential activity in inducing differentiation of human leukemia cells HL-60 *in vitro*, *Proc. Natl. Acad. Sci. USA* 84:2610 (1987).
10. Perlman, K., A. Kutner, J. Prahl, C. Smith, M. Inaba, H.K. Schnoes, and H.F. DeLuca, 24 Homologated 1,25-dihydroxyvitamin D_3 compounds: separation of calcium and cell differentiation activities, *Biochemistry* 29:190 (1990).
11. Zhou, J.Y., A.W. Norman, M. Lubbert, E.D. Collins, M.R. Uskokovic, and H.P. Koeffler, Novel vitamin D analogs that modulate leukemic cell growth and differentiation with little effect on either intestinal calcium absorption or bone calcium mobilization, *Blood* 74:82 (1989).
12. Zhou, J.Y., A.W. Norman, D.-L. Chen, G.-W. Sun, M. Uskokovic, and H.P. Koeffler, 1,25-Dihydroxy-16-ene-23-yne-vitamin D_3 prolongs survival time of leukemic mice, *Proc. Natl. Acad. Sci. USA* 87:3929 (1990).
13. Abe, J., M. Morikawa, K. Miyamoto, S. Kaiho, M. Fukushima, C. Miyaura, E. Abe, T. Suda, and Y. Nishii, Synthetic analogues of vitamin D_3 with an oxygen atom in the side chain skeleton, *FEBS Lett.* 226:58 (1987).
14. Wang, Y.-Z., H. Li, M.E. Bruns, M. Uskokovic, G.A. Truitt, R. Horst, T. Reinhardt, and S. Christakos, Effect of 1,25,28-trihydroxyvitamin D_2 and 1,24,25-trihydroxyvitamin D_3 on intestinal calbindin-D_{9k} mRNA and protein: is there a correlation with intestinal calcium transport?, *J. Bone Min. Res.* 8:1483 (1993).
15. Colston, K., M.J. Colston, and D. Feldman, 1,25-Dihydroxyvitamin D_3 and malignant melanoma: the presence of receptors and inhibition of cell growth in culture, *Endocrinology* 108:1083 (1981).
16. Frampton, R.J., S.A. Omond, and J.A. Eisman, Inhibition of human cancer cell growth by 1,25-dihydroxyvitamin D_3 metabolites, *Cancer Res.* 43:4443 (1983).
17. Eisman, J.A., T.J. Martin, I. MacIntyre, and J.M. Moseley, 1,25-Dihydroxyvitamin D_3 receptors in breast cancer cells, *Lancet* ii:1335 (1979).

18. Brehier, A. and M. Thomassett, Human colon cell line HT-29: characterization of 1,25-dihydroxyvitamin D_3 receptor and induction of differentiation by the hormone, *J. Steroid Biochem.* 29:265 (1988).
19. Colston, K.W., U. Berger, and R. C. Coombes, Possible role for vitamin D in controlling breast cancer cell proliferation, *Lancet* i:188 (1989).
20. Colston, K.W., A.G. Mackay, S.Y. James, L. Binderup, S. Chander, and R.C. Coombes, EB 1089: a new vitamin D analogue that inhibits the growth of breast cancer cells *in vivo* and *in vitro*, *Biochem. Pharmacol.* 44:2273 (1992).
21. Abe, J., T. Nakano, Y. Nishii, T. Matsumoto, E. Ogata, and K. Ikeda, A novel vitamin D_3 analog, 22-oxa-1,25-dihydroxyvitamin D_3 inhibits growth of human breast cancer *in vitro* and *in vivo* without causing hypercalcemia, *Endocrinology* 129:832 (1991).

Chapter 13

Some Aspects of Vitamin E Related to Humans and Breast Cancer Prevention

NIKOLAY V. DIMITROV, RUI-QIN PAN,
JAN BAUER and THOMAS I. JONES

I. Introduction

Vitamin E is widely recognized as a lipotrophic antioxidant which plays an important role in the chain of events that protect biological membranes against injury from a variety of oxidant agents.[1-3] Free radicals provide strongly reactive oxygen in the tissues and can be formed endogenously by normal metabolic processes or exogenously by factors external to the body. These chemical reactions lead directly to cell and tissue damage and are thus thought to initiate or promote neoplasia. It has been suggested that vitamin E can inhibit this process.[2-4] The biological action of vitamin E and the correct evaluation of nutritional status of this micronutrient depend substantially on its kinetics and dynamics in body tissues. Besides the well-characterized function of vitamin E as an antioxidant, other roles such as a regulator of membrane fluidity,[5] a membrane stabilizer[6] and inhibitor of 5-lipoxygenase activity[7] have been demonstrated. Although the exact mechanism of its action is not fully understood, there is some evidence for participation of vitamin E in regulation of growth differentiation and transformation of mammalian cells.[8,9] All these biological activities could be related to the carcinogenic process, and results from animal studies have demonstrated the anticarcinogenic effect of vitamin E in tumors induced by various initiators and promoters.[10,11]

All biological activities of vitamin E are related to cellular function, and the presence of tissue vitamin E is essential to all hypotheses on the biological action of this micronutrient. It is generally accepted that the vitamin must be delivered in sufficient quantity to all vulnerable biomembranes at a sufficient rate and should be retained and maintained at the required organ site. Thus the tissue biokinetics of vitamin E are important for its biological action. During the last four decades a great deal of effort has been made to elucidate the mechanisms by which vitamin E is accumulated in the tissue and its release into the

---

Nikolay V. Dimitrov and Thomas I. Jones ● Department of Medicine, Michigan State University, East Lansing, Michigan; Rui-Qin Pan ● Department of Surgery, China-Japan Friendship Hospital, Beijing, China; Jan Bauer ● Department of Oncology, Charles University, Prague, Czech Republic

Diet and Breast Cancer, Edited under the auspices of the
American Institute for Cancer Research, Plenum Press, New York, 1994

circulation or the surrounding body structures. Attempts have also been made to correlate the amount of the circulating tocopherol with that stored in different organs.[12-14] Although our knowledge about the tissue kinetics and dynamics of vitamin E in animals is substantial, information regarding human tissue storage and the biokinetics of vitamin E is limited.

A relationship between vitamin E and histological changes of breast tissue was suggested in the late sixties and was revived by London and associates. They first reported the benefits from vitamin E in fibrocystic disease and later contradicted their findings in an additional study.[15,16] Since that time several mastalgia clinics have been using vitamin E for control of breast pain in premenopausal women with fibrocystic changes of the breast. The beneficial effect of this treatment was considered as minimal or questionable. However, this negative clinical attempt did not stop the interest of the investigators because the results of some animal and epidemiologic studies indicated the preventive potential of vitamin E against breast cancer.[10,11,17] Storage of vitamin E in the ductal system of the breast and surrounding tissue (fat) is important if the antioxidant effect of alpha-tocopherol is to be utilized for preventive purposes.[14] Although some animal studies indicate a linear relationship between plasma and tissue concentrations of vitamin E in animals, this relationship has not been clearly demonstrated in humans. The tocopherols in the blood are only a tiny part of the total tocopherols in the human body, and the tissue concentrations may vary depending on type and localization. It has been estimated that about 90% of the total body content of tocopherol is found in the adipose tissue.[18] Since vitamin E is considered a potential chemopreventive agent against breast cancer, more information about tissue concentrations of vitamin E in the breast and other storage sites (fat) in the body is needed. We thus designed a pilot study to investigate the adipose tissue concentration of vitamin E in the breast as related to other fat deposits and the ductal system.

The breast is divided into four compartments: skin, ductal system, vascular system, and lymphatics and fat tissue mixed with stromal supportive elements. Interactions between the ductal system and the surrounding fat and stromal tissue have been demonstrated during the development and progression of breast cancer. The role of vitamin E and the relationship between adipose tissue storage and the other structures are not well understood. Studies on mastectomy samples from patients with breast cancer indicate that vitamin E concentrations in the adipose tissue are similar in all four quadrants of the breast, regardless of the tumor location.[19] Since the information about the storage of vitamin E in breast tissue is limited, we undertook pilot studies related to vitamin E concentrations in breast adipose tissue and ducts in tissue samples from normal healthy individuals and patients with breast cancer. The effect of short supplementation with pharmacological doses of vitamin E on adipose tissue deposition and milk was also studied.

II. Methods

A. Sample Collection

Plasma — venous blood was collected from normal healthy individuals before and after oral administration of vitamin E.

Adipose tissue — was collected by surgical biopsy, from adipose tissue suction material obtained during breast reduction procedures and by needle biopsy using the modified method of Hirsch *et al.*[20]

Breast ductal tissue — the ducts were separated from the adipose tissue of the breast by dissection using a thin scalpel.

Milk — was collected from the breast by suction pump before and after oral administration of vitamin E.

B. Analytical Method for Alpha-Tocopherol

Plasma and tissue concentrations of alpha-tocopherol were determined by the method of Nierenberg.[21,22] The adipose tissue preparation for measuring alpha-tocopherol by HPLC was made by the modified method of Kayden.[18]

C. Vitamin E Administration

Vitamin E was given in capsules containing alpha-tocopheryl acetate 400 IU. The patients were taking 1200 IU as a single daily dose for five days before the day of surgery. Female volunteers who were breast feeding their babies took alpha-tocopheryl acetate 800 IU for four days. Plasma and milk alpha-tocopherol concentrations were measured before and after administration of vitamin E.

Informed consent approved by IRB was obtained from the subjects participating in the study.

III. Results

In order to evaluate the distribution of vitamin E in human adipose tissue we collected samples from the right and left upper quadrants of the abdomen and left axilla. The results of these studies are presented in Table 1 and indicate similar distribution in all sample sites. Similar adipose tissue distribution was observed when samples from the breast and axillary fat pad from the same female subject were collected (Table 2). Interindividual differences in the alpha-tocopherol concentrations were noticed again among the participants.

Adipose tissue and plasma concentrations of alpha-tocopherol before and after supplementation with vitamin E for five days are presented in Figure 1. While the plasma concentrations doubled the pretreatment level, the adipose tissue concentrations increased less than 50%. This indicated that increases of plasma and adipose tissue concentrations of alpha-tocopherol after supplementation with pharmacological doses are disproportional.

Table 1. Distribution of Adipose Tissue Alpha-tocopherol in Females[a]

Age	RUQ[b] (n = 10)	LUQ[b] (n = 6)	Left Axilla (n = 7)	P
38-69	137.6 ± 39.4	125.3 ± 33.4	122.8 ± 32.1	NS

[a] Alpha-tocopherol μg/g tissue.
[b] RUQ = right upper quadrant of abdomen; LUQ = left upper quadrant of abdomen.

Table 2. Adipose Tissue Alpha-tocopherol from the Same Female Subjects[a]

Subject	Left Breast	Left Axilla
1	123.6	115.1
2	87.1	96.2
3	139.9	134.2

[a] Alpha-tocopherol μg/g tissue.

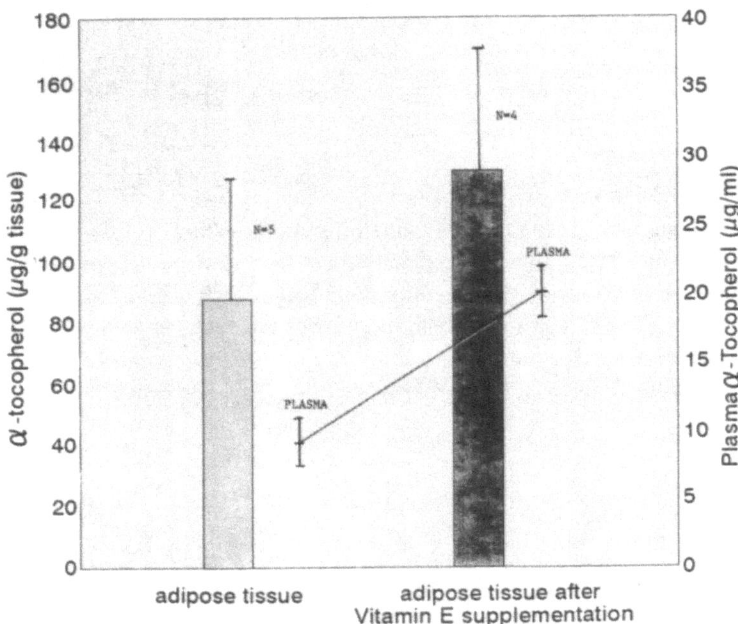

Figure 1. Plasma and adipose tissue concentrations of alpha-tocopherol before and after daily supplementation with all-rac-alpha-tocopherol 1200 IU for five consecutive days.

In order to establish the relationship between concentrations of alpha- and gamma-tocopherol in the adipose tissue and collecting ducts of the breast, we measured separately the tocopherol levels in each tissue. Due to technical problems we were not able to measure the plasma concentrations in this group of experiments. The results presented in Figure 2 indicate that there is a sufficient amount of alpha- and gamma-tocopherol in the collecting ducts which are the main target for oxidative injury in the breast during development of cancer. The concentration of the alpha- and gamma-tocopherol in the ducts is less than that of adipose tissue which is considered as a storage site for the breast. The ratio of alpha- to gamma-tocopherol in the ducts and the adipose tissue is similar. Gamma-tocopherol represents less than 30% of the total tocopherol measured in the breast.

Milk production is a function of the breast glandular system which is composed of lobular and ductal tissue. If vitamin E is found in the collecting duct tissue, the milk as a secretory function of this organ should contain alpha-tocopherol corresponding to the accumulation of the micronutrient in the ductal system. We designed a study to examine the concentrations of alpha-tocopherol in women with lactating breasts. This was done at the end of the breast feeding period before discontinuation of lactation. The results from one female subject are presented in Table 3. The plasma concentrations rose substantially as expected. The milk concentration at the end of the vitamin E dosing increased almost three times suggesting a relationship between the plasma level and milk concentrations. The kinetic pattern of the milk observed in this subject indicates rapid increase of tocopherol concentrations after the first day of supplementation.

Since breast cancer represents a malignancy with enormous biological and morphological heterogeneity, we decided to measure the intratumoral concentrations of alpha-tocopherol as related to patients' hormonal dependence. Table 4 shows the results from measurements of alpha-tocopherol in breast tumor tissues taken from different patients. We did not measure

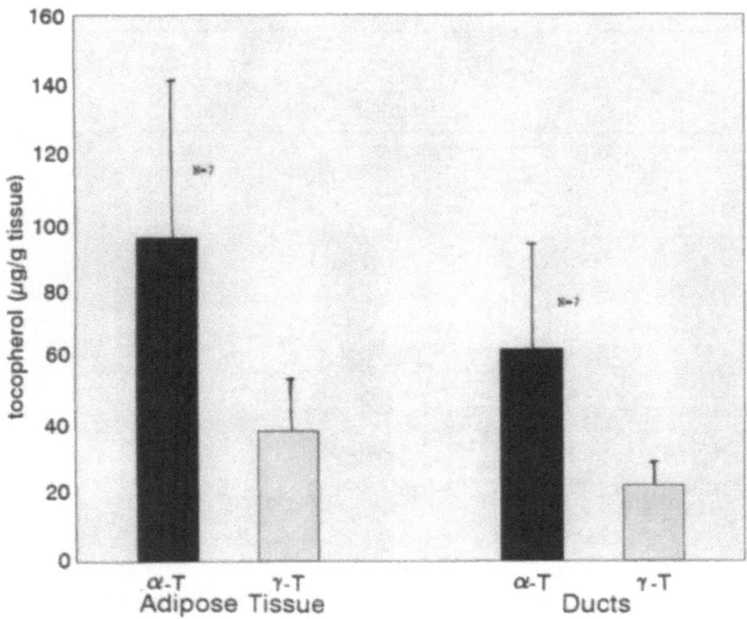

Figure 2. Concentrations of alpha- and gamma- tocopherols in adipose tissue and the collecting ducts from the same breast in seven female subjects.

plasma levels for correlation with the concentrations in the tumor. There are two distinctive groups of patients with different estrogen receptor (ER) status and different patterns of alpha-tocopherol tumor concentrations. Tumors with the ER positive marker have higher concentrations compared to the ER negative tumors. This pattern also correlates with the differentiation of the tumor. Poorly differentiated tumors had lower concentrations of alpha-tocopherol compared to those with well and moderately differentiated patterns.

IV. Discussion

Evaluation of long term vitamin E status in humans using adipose measurement has advantages over determination of plasma concentrations. Plasma levels of alpha-tocopherol change very rapidly in humans following a change in vitamin E intake, reaching a higher steady state in a few days.[23,24] Distribution of adipose tissue vitamin E at different sites is important for eventual supply to the surrounding structures. Although the exact mechanism of vitamin E uptake and retention by adipose tissue is unknown, there is strong evidence that lipoprotein lipase plays an important role in the distribution of tocopherol from chylomicrons to the adipose tissue.[18] Thus the distribution of tocopherol in different locations of adipose tissue is expected to be similar. The results from this pilot study indicate that adipose tissue alpha-tocopherol concentrations in samples from abdominal wall and the axillary area are similar (Table 1). The adipose tissue concentrations in the breast adipose tissue fall in the same range (Table 2). Several other studies done on human autopsy material showed no significant differences in tocopherol concentrations of the adipose tissue taken from different sites of the body.[25,26] However, samples from one subject, taken on the same day at different anatomical sites, showed higher levels at the left and right waist sites compared to

Table 3. Kinetics of Milk Tocopherol Concentrations After Supplementation of Vitamin E[a]

	Plasma		Milk	
Time (hrs)	alpha-T[b]	gamma-T	alpha-T	gamma-T
0	9.2	2.4	3.4	1.6
12			10.4	4.1
36	19.6	0.55	17.2	6.2
60	17.6	0.99	14.3	4.0
84	-	-	12.2	3.1
156	-	-	9.1	1.6
240	9.4	1.9	5.4	1.1
324	-	-	2.7	0.8

[a]   D-alpha-tocopheryl acetate 800 IU for four days.
[b]   Alpha-T = alpha-tocopherol.

Table 4. Tocopherol Concentrations in Estrogen Receptor Positive or Negative Breast Tumor Tissue

Tumor Tissue α-Tocopherol[a]	ER Status	PD	WD	MWD
162	+		+	
52	+		+	
141	+		+	
429	+		+	
183	+			+
122	+			+
137	+		+	
39	-	+		
52	-	+		
30	-	+		
111	-	+		
32	-	+		

[a]   µg/g tissue.
[b]   ER-estrogen receptor; PD-poorly differentiated; WD-well differentiated; MWD-moderately well differentiated.

gluteal adipose tissue.[27] Since the interindividual variations appear to be in a wide range (Table 1), a single experiment cannot determine the real pattern of distribution.

Sampling different quadrants of the breast also revealed similar concentrations. Similar findings were reported by Rautalahti *et al.*[19] in studies of peritumoral adipose tissue obtained from different quadrants of the breast. If carcinogenesis occurs in the glandular epithelium, then the main target for preventive measures should be the ductal system of the breast. The results from measurement of ductal concentrations of alpha-tocopherol (Figure 2) revealed that the micronutrient is present in the ducts in substantial amounts but which are lower than in the surrounding adipose tissue. It is unknown whether supplementation with pharmacological doses of vitamin E will increase the concentrations in both the ductal system and surrounding adipose tissue. To our knowledge no kinetic studies have been done in humans related to adipose tissue and vitamin E supplementation. The kinetic studies of repletion and depletion in animals suggest that adipose tissue alpha-tocopherol turnover is slower than in other tissues.[12,13]

Since the ducts of the breast are considered as excretory glands, the concentration of tocopherol in the milk may reflect the level of tocopherol in the glandular tissue. It is generally accepted that vitamin E plays an important role in preventing oxidation of the milk.[28] If increased doses of vitamin E will result in an increase of milk tocopherol concentrations, then the micronutrient has been successfully delivered to the ducts which are also the target of oxidative damage as part of the carcinogenic process. We have measured milk tocopherol concentration in one female subject who was involved in breast feeding (Table 4). The amount of vitamin E present in the ducts of the breast appeared to be sufficient to protect the milk from oxidation and to protect the ducts from oxidative damage. Since there is evidence that mammary dysplasia is associated with lipid peroxidation, vitamin E should be considered as a preventive agent, preferably in selected cases indicating dysplastic changes.[29] Although the precise biochemical functions of tocopherol are as yet unknown, accumulating evidence indicates that it presumably functions as an antioxidant *in vivo*.[2,30] It has been suggested that tocopherol in the mammary gland may function as an antioxidant to protect the store of unsaturated fatty acids from being oxidized.[28] This may explain why the major portion of tocopherol (about 96%) is stored in the adipose tissue.

While the tocopherol concentrations in normal breast (fat and ducts) may be needed as protectors against oxidative damage, the concentrations of tocopherol in breast cancer tissue may play a different role. The results from our studies indicate that breast cancer tissues which possess estrogen receptors contain higher concentrations of tocopherol (Table 3). To the contrary, estrogen receptor negative tumor showed lower concentrations of both alpha- and gamma-tocopherols. The estrogen receptor negative tumors usually respond well to administration of chemotherapy, whereas the estrogen positive receptor tumors show lower response rates.[31] The action of adriamycin (a potent anticancer agent) against breast cancer cells is based on the release of free radicals which damage the cells.[32] If the concentration of alpha- and gamma-tocopherols in the cancer cells is high, then the released free radicals will be trapped by the tocopherol and will not be able to exercise the anticancer effect. Thus, high levels of tocopherol in the breast tumor tissue may increase the resistance to chemotherapy.[33] This area will need further research in order to evaluate the significance of vitamin E in breast cancer tissue.

Our preliminary work on deposition of vitamin E in the breast (ducts and adipose tissue) dealt only with one antioxidant, but evidence is accumulating that other antioxidants may act synergistically in preventing tissue oxidative injury. A careful evaluation of micronutrients recommended as possible chemopreventive agents against breast cancer alone or in combination may help to avoid undue anticipation or disappointment regarding interpretation of the results from early intervention trials.

V. Summary

The biological activities of vitamin E are related to the cellular functions and presence of sufficient tissue concentrations of this micronutrient. Most of the stored vitamin E is in the adipose tissue where it appears to be distributed equally. The breast adipose tissue has similar vitamin E concentrations as other parts of the body. The ductal systems also store vitamin E in sufficient concentrations to maintain cellular functions. The milk secreted from the ducts of the breast contains a high concentration of tocopherol.

Whereas the normal breast tissue presumably utilizes vitamin E as an antioxidant, tumor tissue appears to handle vitamin E differently. Breast tumors possessing estrogen negative receptors and having poor histological differentiation have lower concentrations of vitamin E than tumors with positive estrogen receptors and well differentiated histology.

Since vitamin E is considered the principal, if not sole, chain-breaking lipophilic antioxidant in plasma and tissue, its role as a potential chemopreventive agent in breast cancer should be further investigated. The combination of vitamin E with other cancer chemopreventive agents appears to be a reasonable procedure.

References

1. Machlin, L.J., A comprehensive treatise, *in: Vitamin E*, L.J. Machlin, ed., Marcel Dekker, New York (1980).
2. Burton, G.W. and K.U. Ingold, Vitamin E as an *in vitro* and *vivo* antioxidant, *Ann. NY Acad. Sci.* 570:7 (1989).
3. Kneckt, P., Role of vitamin E in prophylaxis of cancer, *Ann. Med.* 23:3 (1991).
4. Pascoe, C.A. and D.J. Reed, Vitamin E protection against chemical induced cell injury, *Arch. Biochem. Biophys.* 256:159 (1987).
5. Niki, E., Inhibition of oxidation of liposomal- and bio-membranes by vitamin E, *in: Clinical and Nutritional Aspects of Vitamin E*, O. Hayaishi and M. Mino, eds., Elsevier, Amsterdam (1987).
6. Urano, S., K. Yano, and M. Matsuo, Membrane stabilizing effect of vitamin E: effect of alpha-tocopherol and its model compounds, *Biochem. Biophys. Res. Commun.* 150:469 (1988).
7. Redanna, P., J. Whelen, J.R. Burgess, M.L. Eskew, G. Hildenbrand, A. Zarkower, R.W. Scholz, and C.C. Reddy, The role of vitamin E and selenium on arachidonic acid metabolism by way of the lipoxygenase pathway, *Ann. NY Acad. Sci.* 570:136 (1989).
8. Boscoboinic, D., A. Szewiszyk, C. Hensey, and A. Azzi, Inhibition of cell proliferation by alpha-tocopherol, *J. Biol. Chem.* 6188 (1991).
9. Huang, N., B. Lineberger, and M. Steiner, Alpha-tocopherol, a potent modulator of endothelial cell function, *Thromb. Res.* 50:547 (1988).
10. Ip, C., Dietary vitamin E intake and mammary carcinogenesis in rats, *Carcinogenesis* 3:1453 (1982).
11. Wang, Y.-M., M. Purewal, B. Nixon, D.-H. Li, and D. Soltysiak-Pawluczuk, Vitamin E and cancer prevention in an animal model, *Ann. NY Acad. Sci.* 570:383 (1989).
12. Biery, J., Kinetics of tissue alpha-tocopherol depletion and repletion, *Ann. NY Acad. Sci.* 203:181 (1972).
13. Machlin, L.J., J. Keating, J. Nelson, M. Brin, R. Filipski, and O.N. Miller, Availability of adipose tissue tocopherol in the guinea pig, *J. Nutr.* 109:105 (1979).
14. Ingold, K.U., G.W. Burton, D.O. Foster, L. Hughes, D.A. Lindsay, and A. Webb, Biokinetics of and discrimination between dietary RRR- and SRR-α-tocopherols in the male rat, *Lipids* 22:163 (1987).
15. London, R.S., L. Murphy, and K.E. Kitlowski, Hypothesis: breast cancer prevention by supplemental vitamin E, *J. Am. Coll. Nutr.* 4:559 (1985).
16. London, R.S., G.S. Sundaram, L. Murphy, S. Manimekalai, M. Reynolds, and P.J. Goldstein, The effect of vitamin E on mammary dysplasia: a double-blind study, *Obstet. Gynecol.* 65:104 (1985).
17. Pryor, W.A., Vitamin E: The status of current research and suggestions for future studies, *Ann. NY Acad. Sci.* 570:400 (1989).
18. Kayden, H.J., Tocopherol content of adipose tissue from vitamin E deficient humans, *in: Biology of Vitamin E*, R. Porter and R. Whelan, eds., CIBA Foundation Symposium 101, Pitman Books, London (1983).

19. Rautalahti, M., D. Albanes, L. Hyvonen, V. Piironen, and M. Heinonen, Effect of sampling site on retinol, carotenoid, tocopherol and tocotrienol concentrations of adipose tissue of human breast with cancer, *Ann. Nutr. Metab.* 34:37 (1990).

20. Hirsch, J., J.W. Farquhar, E.H. Ahrens, M.L. Peterson, and W.M. Stoffel, Studies of adipose tissue in man, *Am. J. Clin. Nutr.* 8:499 (1960).

21. Nierenberg, D.W. and D.C. Lester, Determination of vitamins A and E in serum and plasma using a simplified clarification method and high-performance liquid chromatography, *J. Chromatogr.* 345:275 (1985).

22. Nierenberg, D.W. and S.L. Nann, A method for determining concentrations of retinol, tocopherol and five carotenoids in human plasma and tissue samples, *Am. J. Clin. Nutr.* 56:417 (1992).

23. Baker, H., O. Frank, B. DeAngelis, and S. Feingold, Plasma tocopherol in man at various times after ingesting free or acetylated tocopherol, *Nutr. Rep. Int.* 21:531 (1980).

24. Dimitrov, N.V., C. Meyer, D. Gilliland, M. Ruppenthal, W. Chenoweth, and W. Malone, Plasma tocopherol concentrations in response to supplemental vitamin E, *Am. J. Clin. Nutr.* 53:723 (1991).

25. McMasters, V., J.K. Lewis, L.W. Kinsell, J. van der Veen, and H.S. Olcott, Effect of supplementing the diet of man with tocopherol on the tocopherol levels of adipose tissue and plasma, *Am. J. Clin. Nutr.* 17:357 (1965).

26. Imaichi, K., S. Cox, L.W. Kinsell, M. Schelstraete, and H.S. Olcott, Tocopherol content of human adipose tissue, *Am. J. Clin. Nutr.* 16:347 (1965).

27. Handelman, G.J., W.L. Epstein, L.J. Machlin, F.J.G.M. vanKuijk, and E.A. Dratz, Biopsy method for human adipose with vitamin E and lipid measurements, *Lipids* 23:598 (1988).

28. Hidirogiou, M., Mammary transfer of vitamin E in dairy cows, *J. Dairy Sci.* 72:1067 (1989).

29. Boyd, N.F. and V. McGuire, Evidence of lipid peroxidation in premenopausal women with mammographic dysplasia, *Cancer Lett.* 50:31 (1990).

30. McCay, P.B., Vitamin E: interactions with free radicals and ascorbate, *Ann. Rev. Nutr.* 5:323 (1985).

31. Kardinal, C.G., Chemotherapy of breast cancer, in: *The Chemotherapy Source Book*, M.C. Perry, ed., Williams & Wilkins, Baltimore (1992).

32. Myers, C.E., W.D. McGuire, I. Ifrim, R.H. Liss, K. Grotzinger, and R.C. Young, Adriamycin: the role of lipid peroxidation in cardiac toxicity and tumor response, *Science* 197:165 (1977).

33. Shinozawa, S., Y. Gomita, and Y. Araki, Effect of high dose alpha-tocopherol and alpha-tocopherol acetate pretreatment on adriamycin (doxorubicin) induced toxicity and tissue distribution, *Phys. Chem. Phys. Med. NMR* 20:329 (1988).

Poster Abstracts

A Case Control Study of Chili Pepper Consumption and Gastric Cancer in Mexico

L. LOPEZ-CARRILLO, R. DUBROW, M. HERNANDEZ-AVILA, J. PAPAQUI-HERNANDEZ and C. FERNÁNDEZ-ORTEGA

National Institute of Public Health, Mexico City, Mexico

Hot chili peppers are heavily consumed in Mexico. Animal and *in vitro* studies indicate that capsaicin, the hot-tasting component of chili peppers, may be carcinogenic. Gastric cancer accounts for 10% of cancer deaths in Mexico. This is the first epidemiologic study to evaluate the relationship between chili pepper consumption and gastric cancer risk.

A population-based case-control study was conducted in the Mexico City metropolitan area. Two-hundred and twenty newly diagnosed gastric adenocarcinoma cases were compared with 752 controls randomly selected from the general population. Interviewers administered a structured questionnaire that included a semi-quantitative food frequency inventory adapted to the Mexican diet, as well as specific questions about chili pepper consumption.

After adjustment for age and sex, chili pepper consumption was found to be a strong risk factor for gastric cancer (odds ratio = 5.15; 95 percent confidence interval, 2.56 to 10.37). A highly significant trend of increasing risk with increasing self-rated level of consumption (none, low, medium, high) was observed ($p = 5 \times 10^{-14}$). The odds ratio for persons who rated their level of consumption as high was 17.08 (95% confidence interval, 7.77 to 37.53) compared with non-consumers. Multivariable adjustment for age, sex, total calories, socioeconomic status, history of peptic ulcer, and consumption of fruits, vegetables, processed meats, beans, salt, alcohol, and cigarettes did not alter the magnitude of the association.

Chili pepper consumption thus appears to be a strong risk factor for gastric cancer in Mexico. Further studies are needed to confirm or refute this finding.

Effect of Selenium on Natural Killer Cell Activity in Humans and Mice

L. KIREMIDJIAN-SCHUMACHER, M. ROY, H.I. WISHE, M.W. COHEN and G. STOTZKY

New York University Dental Center and Department of Biology, New York, NY

This study examined the effect of dietary supplementation with selenium (Se; as sodium selenite) on the ability of human peripheral blood and murine splenic natural killer cells to destroy tumor cells. The human participants in the study were randomized for age, sex, weight, height, and nutritional habits and given selenite (200 µg/day) or placebo tablets for 8 weeks; all participants had a selenium replete status as indicated by their plasma Se levels prior to supplementation. For the studies on the mouse model, 6 week old C57Bl/6J male mice were maintained on a Se-normal (0.2 ppm) or Se-supplemented (2.0 ppm) torula yeast-based diet for 8 weeks. The data indicated that the supplementation regimen resulted in 82.3% (human) and 39.2% (mouse) increase in natural killer cell activity. This apparently was related to the ability of the nutrient to enhance the expression of receptors for the growth regulatory lymphokine interleukin 2 and, consequently, the rate of cell proliferation and differentiation into cytotoxic cells. The supplementation regimen did not produce significant changes in the plasma Se levels of the human participants. The results indicated that the immunoenhancing effects of selenium in humans require supplementation above the replete levels produced by normal dietary intake.

Supported by the American Institute for Cancer Research grant 91A01.

Effect of Selenium on the Proliferation of Human Lymphocytes and the Expression of Il$_2$-R

M. ROY, L. KIREMIDIIAN-SCHUMACHER, H.I. WISHE, M.W. COHEN and G. STOTZKY

New York University Dental Center and Department of Biology, New York, NY

Selenium (Se) is an essential nutritional factor that was shown previously by us to alter the expression of the high affinity interleukin 2 receptor (Il$_2$-R) and its subunits, cell proliferation, and the clonal expansion of cytotoxic T lymphocytes in mice. This study shows that dietary supplementation of Se-replete humans with 200 µg/day of sodium selenite for 8 weeks, or *in vitro* supplementation with 1 x 10^{-7} M Se (as sodium selenite), results in a significant augmentation of the ability of peripheral blood lymphocytes to respond to stimulation with 1 µg/ml phytohemagglutinin or alloantigen (mixed lymphocyte reaction), and to express high affinity Il$_2$-R on their surfaces. There was a clear correlation between enhanced nuclear DNA ^{3}H-thymidine incorporation, preceded by enhanced high affinity Il$_2$-R expression, and supplementation with Se. It appears that supplementation with Se in humans can modulate T lymphocyte mediated immune responses that depend on signals generated in the interaction of interleukin 2 with Il$_2$-R.

Supported by the American Institute for Cancer Research grant 91A01.

Activation of Protein Kinase C by Lipoxygenase Metabolites

R.A. HAGERMAN and M.F. LOCNISKAR

Nutritional Sciences, University of Texas, Austin, TX

Dietary lipids modulate events which occur during the promotion stage of the initiation-promotion skin carcinogenesis model. Metabolites of linoleic and arachidonic acid have been shown to be inv lved in events of promotion induced by 12-*O*-tetradecanoylphorbol-13-acetate (TPA), an ac for of protein kinase C (PKC). We and others have shown that PKC can be activated *in vitro* by fatty acids. However, the effect of lipids on the *in vivo* distribution of PKC is not well understood. Therefore, experiments were designed to address this question. First, we established the time course and cellular location of incorporation of exogenous fatty acids. Second, we assessed the *in vivo* modulation of PKC distribution by fatty acids. In these experiments, PGE_2 and $PGF_{2\alpha}$ were chosen because they are known to modulate epidermal TPA-induced tumor promotion. The lipoxygenase products of linoleic acid (13(S)-HODE) and arachidonic acid (12(S)-HETE) were selected because we have found that they are the primary metabolites of lipoxygenase in mouse epidermis.

Keratinocytes were isolated from newborn Sencar mice by trypsinization. After plating, cells were grown in defined medium (SPRD-111) and incubated with 10^{-9} M ^{3}H-12(S)-HETE. Treated cells were washed and total lipids extracted. Phospholipids were separated by thin layer chromatography and extracted from silica. Half of the recovered lipid was counted using liquid scintillation counting and the other half assayed for phosphorus content. Data were expressed as a percentage of total for each time point. The majority of labelled lipid was incorporated at 6 hours into phosphatidylcholine (PC) (42%) and phosphatidylethanolamine (PE) (50%), but it was found primarily in PE by 18 hours (PC 12%; PE 80%). Only minor amounts of lipid were incorporated into phosphatidylinositol (PI) or phosphatidylserine (PS) (PI 4.2%; PS 4.6%). To determine the molecular location of fatty acid incorporation, we examined whether ^{3}H-12(S)-HETE could be released via activation of phospholipase A_2 by TPA. After 6 hours incubation with label, 1.0 µg/ml TPA was added to cultures. Medium was collected at 6 hours and counted by liquid scintillation. At 6 hours, 260% more ^{3}H-12-HETE was released after TPA treatment compared to acetone (vehicle).

The activation of PKC by fatty acids was assessed by incubating keratinocytes with 10^{-8} M of a compound for 18 hours. Cytosolic (C) and particulate (P) fractions of PKC were separated and activity determined by ^{32}P phosphorylation of histone III-S. Results were expressed as a percentage of total activity. In vehicle (ethanol) treated cells, PKC distribution was 29 (P):71 (C). Incubation with 12(S)-HETE caused translocation of PKC to the membrane (68:32), but pre-treatment of cells with 13(S)-HODE prevented this shift (33:67). PGE_2 also caused translocation (75:26); however, $PGF_{2\alpha}$ did not (36:63).

In summary, fatty acid metabolites incorporated into phospholipids can cause translocation of PKC, and may interact with each other to inhibit this effect.

Supported by National Institutes of Health grant CA 46886 and by U.S. Department of Agriculture.

Alterations in Epithelial-Extracellular Matrix Interactions in the Breast by Chemopreventive Agents DFMO and RA

P.J. SCHEDIN, M. KAECK, R. STRANGE,
M. SINGH and H.J. THOMPSON

AMC Cancer Research Center, Denver, CO

The effect of retinyl acetate (RA) and D,L-2-difluoromethylornithine (DFMO) fed alone or in combination on mammary carcinogenesis was investigated in female Sprague-Dawley rats injected i.p. with 50 mg 1-methyl-1-nitrosourea (MNU)/kg at 50 days of age. Beginning 7 days post MNU, rats were fed a lab chow diet into which 1 mmol retinyl acetate/kg and/or 1% DFMO (w/w) were added. Both DFMO and RA inhibited mammary carcinogenesis, but the combination of DFMO and RA was most effective in reducing tumor multiplicity and increasing latency. Histological evaluation of mammary glands from DFMO/RA-treated animals demonstrated both a reduction in the number of ducts and decreased size of alveolar units. An increase in mammary gland extracellular matrix was also observed. The reduced gland complexity observed was consistent with these agents acting either directly to inhibit proliferation or indirectly by blocking differentiation.

The proliferative and differentiative status of the rat mammary gland, in part, determine the susceptibility of the gland to tumor promotion/progression. Exogenous and endogenous mammotrophic hormones act as tumor promoters in the breast. Thus, the effect of DFMO or RA on 17-β-estradiol and progesterone-stimulated proliferation and differentiation also was assessed. Neither DFMO nor RA exerted an effect on hormone-driven cell proliferation. However, expression of β-casein and WAP mRNAs was greatly reduced. Combined DFMO and RA treatment resulted in greatest accumulation of extracellular matrix and the largest reduction in the epithelial component of the gland. These changes were correlated with elevated expression of tissue transglutaminase mRNA, a gene associated with apoptotic cell death. Based on our observations, we propose that DFMO and RA interfere with mammotrophic hormone stimulation in the breast by altering epithelial-mesenchymal interactions which, in turn, shift the balance among cell proliferation, differentiation and death.

Cervical Cancer and Diet:
A Case Control Study in Mexico

E. LAZCANO, L. LOPEZ-CARRILLO, I. ROMIEU,
P. ALONSO, and M. HERNANDEZ-AVILA

National Institute of Public Health, Mexico City, Mexico

Cervico-uterine neoplasm (CU CA) is the leading cause of death for cancer among Mexican females. According to the Mexican health figures, there were 400 deaths from CU CA in 1990, with a corresponding mortality rate of 7.6/100,000 women. Currently, the CU CA incidence rate in females over 25 years old is estimated to be 115 per 100,000.

Our objective was to evaluate and to quantity the possible association between cervico-uterine cancer and: 1) food and vitamin consumption, 2) reproductive factors, and 3) factors associated with life style (sociodemographic characteristics, tobacco use).

A case-control study was carried out in Mexico City from September 1990 to May 1993. 513 incident cases of cervical neoplasm were registered in 8 hospitals: Two were from the Ministry of Health, four were Social Security hospitals, and two were private hospitals. From a random sample of homes in the 16 political districts of Mexico City and 7 suburban municipalities, 1007 population controls were collected. The response rate of participation in the study was above 84.2%. The interview included questions about: 1) Risk CU CA factors; 2) Semi-quantitative food frequency and 3) Clinical, cytological and histological report of cases.

Multivariated adjusted odds ratios were obtained for those variables that resulted in factors related to the incidence of this tumor: age at first sexual intercourse (over 25 years R.M.=0.54 and 95% C.I. 0.32-0.91), Caesarean (R.M.=0.31 and 95% C.I. 0.13-0.69), consumption of green vegetables (in the fourth quartile of consumption R.M.=0.58 and 95% C.I. 0.43-0.79), vitamin A (fourth quartile R.M.=0.58 and 95% C.I. 0.41-0.82) and vitamin C (fourth quartile R.M.=0.61 and 95% C.I. 0.42-0.89). Multiple vaginal births (after the fifth birth R.M.=0.98 and 95% C.I. 1.21-3.25), a history of 2 or more sexual partners (4 or more partners R.M.=4.75 and 95% C.I. 1.92-11.75) increase the risk for the disease.

The principal risk factors for cervico-uterine cancer are early age at first sexual intercourse, multiple vaginal births, and sexual promiscuity. Caesareans, high consumption of green vegetables, or of A and C vitamins have a protective effect. This is relevant information for use in population detection programs for women at high risk of the disease. In relation to food consumption, the study provides an empirical basis for the development of random clinical tests with vitamin supplements.

The Development of Drug Resistance by Tumor Cells *in vitro* is Accompanied by the Development of Sensitivity to Selenite

P.B. CAFFREY and G.D. FRENKEL

Department of Biological Sciences,
Rutgers University, Newark, NJ

We previously demonstrated that cells derived from a drug-resistant human ovarian tumor were more sensitive to selenite than cells derived from a comparable drug-sensitive tumor [Caffrey and Frenkel, *Cancer Res.* 52: 4812 (1992)]. This led us to investigate the hypothesis that the development of drug resistance in tumor cells is accompanied by the development of increased sensitivity to selenite. We examined the effect of selenite on a line of melphalan-resistant (A2780ME) cells which had been developed *in vitro* from a drug-sensitive line (A2780) [Hamilton *et al. Biochem. Pharmacol.*, 34: 2583 (1985)]. We found that the drug-resistant cells exhibited increased sensitivity to the inhibitory effects of selenite on the ability of cells to exclude Trypan Blue dye and to proliferate. As a result, exposure to selenite under conditions in which the number of drug-sensitive cells remained stable resulted in the elimination of more than 99% of the viable drug-resistant cells. The development of melphalan resistance in these cells has been shown to be due to the increase in the level of intracellular glutathione (GSH). To determine whether GSH is also responsible for the increase in selenite sensitivity, we examined the effect of GSH depletion (with buthionine sulfoximine) on the sensitivity of the cells to selenite. We also examined the relative sensitivity of these cells to selenodiglutathione, the product of the reaction of selenite with GSH. The results have demonstrated that the development of drug resistance is accompanied by an increase in sensitivity to selenite and that both of these phenotypic changes are the result of an increase in the level of intracellular GSH.

Supported by grants from the American Institute for Cancer Research and National Institutes of Health.

Dietary Factors in Breast Cancer in Pakistan

F. NIZAMI

Pakistan Medical Research Council,
College of Physicians & Surgeons Pakistan, Karachi, Pakistan

A case control study consisting of 100 females with biopsy proven breast cancer and 100 normal females matched by age was conducted to assess if any dietary factor is responsible for the development of breast cancer. Ninety percent of the cases were over the third decade of life. Anthropometric measurements indicated greater prevalence of obesity in cases than in controls and severity of obesity (fat deposition) was increased in cases along with older age groups. In controls, no correlation of obesity and older age groups was found. Recall of past 24 hours dietary intake indicated 25% of total caloric intake in cases and 18% in controls was from fat. The intake of animal fat (saturated fat) as well as sources of animal protein which are also high in saturated fat was significantly higher in cases than in controls. The breakdown of fat intake in various age groups shows that saturated fat intake was higher in all age groups of cases, while in controls the trend was considerably less in older age groups.

The frequency of different food intakes indicated that the intake of legumes, cruciferous vegetables as well as fruits rich in vitamins A and C which are known to have an anti-carcinogenic effect was significantly lower in cases than in controls. No effect of height as a risk factor for breast cancer was noted. Nutritional factors related to body weight can be a risk factor for the development of breast cancer at any time in adult life.

Effect of Omega-3 Fatty Acids on Membrane Structure and Expression of Membrane Antigens

L.J. JENSKI, G. BOWKER, A.C. DUMAUAL,
L.K. STURDEVANT and W. STILLWELL

*Department of Biology, Indiana University-
Purdue University at Indianapolis, IN*

Diets rich in omega-3 fatty acids, which are abundant in fish oils, are associated with reduced incidence of certain cancers. In addition to affecting eicosanoid hormone production, omega-3 fatty acids play a structural role in biological membranes. We are exploring how omega-3 fatty acid-induced structural changes in tumor and lymphocyte membranes affect immunity to cancer. In this project, mice are fed diets containing fish oil, corn oil, or saturated fat, and the structure and function of tumor and lymphocyte membranes are analyzed; alternatively, tumor cells and lymphocytes are modified with purified lipids *in vitro*, and the analyses are performed. Using a variety of membrane probes (merocyanine 540, laurodan, dansyl-lysine, pyrene), we have found that tumor membranes undergo structural alterations dependent upon the presence of the omega-3 fatty acid docosahexaenoic acid (DHA) in the lipid bilayer. Concurrent with these structural changes are changes in the expression of membrane proteins, such as major histocompatibility complex proteins targeted by the immune system; additionally, T-lymphocyte surface proteins such as Thy-1 are modulated. Although some modification of protein biosynthesis may occur in the presence of omega-3 fatty acids *in vivo*, the experiments performed *in vitro* indicate that membrane-bound omega-3 fatty acids have additional effects, like influencing the conformation of membrane proteins or inducing formation of membrane microdomains with distinct protein composition. We are currently working to distinguish these possible mechanisms. Our results will not only yield basic information about omega-3 fatty acids' role in membrane structure, they will also provide insight into the health benefit of dietary fish oils.

Effects of Dietary Iron on Distribution of Iron in Mammary Gland Cancer, on the Induction of Oxidative DNA Damage and the Genetic Alterations Observed in 1-Methyl-1-Nitrosourea (MNU) Induced Mammary Carcinomas

H.J. THOMPSON, M. SINGH, J. MCGINLEY, A. HAEGELE, J. LU and M. KAECK

AMC Cancer Research Center, Denver, CO

Female Sprague-Dawley rats were injected with 35 mg MNU/kg at 50 days of age and then randomized to test groups fed 35, 350 or 3500 µg Fe/kg as ferrous sulfate. Increasing dietary iron was associated with a biphasic effect on tumor occurrence. In comparison to rats fed 35 µg Fe/g, an enhanced tumorigenic response was observed in those fed 350 µg Fe/g; whereas, feeding 3500 µg Fe/g resulted in a reduction in tumor occurrence. Increasing dietary iron resulted in detection of stainable iron within the mammary gland. In non-tumor bearing rats, staining was greater in the epithelial cells of mammary lobules than ducts. Stainable iron was not detected in mammary gland stromal tissue. Under these conditions, levels of 8-hydroxydeoxyguanosine (8-OHdG) in DNA isolated from the mammary gland were elevated with increasing iron. Antisera against 8-OHdG were used to demonstrate definitive staining in mammary epithelial cells. In animals bearing tumors, distribution of iron within the mammary gland was altered. Iron staining of epithelial cells in mammary carcinomas was minimal whereas that of stromal septa was increased. Uninvolved mammary gland adjacent to the tumor displayed significant alterations in staining. More iron was apparent. Epithelial, connective and adipose components of the gland had detectable iron. 8-OHdG levels in DNA isolated from tumors were not significantly greater than those found in DNA isolated from mammary gland. All mammary cancers were evaluated for mutant Ha-*ras* positivity. The proportion of tumors positive for this mutation was reduced when dietary iron was increased to 350 µg/kg.

Supported by the American Institute for Cancer Research grant 90B55.

Effects of Increasing Levels of Soluble Dietary Fibers on Experimental Colon Carcinogenesis in Rats

D. KRITCHEVSKY, M.W. WEBER and D.M. KLURFELD

Wistar Institute, Philadelphia, PA

It is generally perceived that insoluble fiber, especially wheat bran, reduces the incidence of experimentally-induced colon cancer in rats. The role of soluble fiber in colon carcinogenesis is less well defined. We have examined the effects of increasing concentrations (5, 10, 20%) of dietary pectin and guar gum on incidence of DMH-induced colon cancer in male F344 rats and have attempted to relate our findings to intestinal pH and fecal weight.

Male F344 rats (24/group) were maintained on a commercial laboratory ration and starting at 70 days of age, each rat was given a weekly dose of 1,2-dimethylhydrazine (DMH) (30 mg/kg administered by gavage for 6 weeks). One week after the last treatment all rats were placed on a semipurified diet containing 5, 10 or 20% of either pectin or guar gum. Necropsies were performed on all rats after 26 weeks. Guar gum and pectin at 20% of the diet decreased tumor incidence but not multiplicity. Both fibers decreased cecal and distal colonic pH to a significant degree.

Tumor incidence and multiplicity in a fiber-free group were 71 % and 2.4 ± 0.3, respectively. The decreasing tumor incidence is correlated with decreasing cecal and colonic pH and may be an indication of a mechanism of action of these soluble fibers. Fecal dry weight (g/day) increased significantly with increasing fiber content (p<0.001 for both pectin and guar). Increasing fecal bulk has been suggested as a mechanism by which dietary fiber may reduce risk of colon cancer.

Effects of Prolonged Choline Deficiency and Subsequent Refeeding of Choline on 1,2-*sn*-Diradylglycerol and Protein Kinase C in Rat Liver

K. DA COSTA, S.C. GARNER,
J. CHANG and S.H. ZEISEL

University of North Carolina, Chapel Hill, NC

Rats fed a choline deficient diet develop foci of enzyme-altered hepatocytes with subsequent formation of hepatic tumors. They also develop fatty livers because choline is needed for hepatic secretion of lipoproteins. We have previously reported that 1,2-*sn*-diradylglycerol accumulates in the livers of rats fed a choline deficient diet for 1-27 weeks and that protein kinase C activity in the hepatic plasma membrane is elevated during that time (*J. Biol. Chem.* 268:2100, 1993). In the present study, we examined the changes that occur in rat liver at 52 weeks of choline deficiency, and we determined whether these changes were reversible when choline was returned to the diet of the deficient animals for 1 and 16 weeks. At 52 weeks, the experimental animals had increased 1,2-*sn*-diradylglycerol concentrations in hepatic lipid droplets compared to control animals. Plasma membrane 1,2-*sn*-diradylglycerol levels in the liver did not differ between the two groups, but an age-related increase in membrane 1,2-*sn*-diradylglycerol concentrations was observed. Protein kinase C activity associated with the plasma membrane remained significantly elevated in the deficient livers at 52 weeks. Hepatic foci expressing γ-glutamyltranspeptidase were detected only in the deficient rats (0.83% liver volume), and 15% of these rats had hepato-cellular carcinoma at 1 year on the diet. At 53 weeks (1 week after choline was returned to the deficient group), 1,2-*sn*-diradylglycerol concentrations in the hepatic lipid droplets had dropped to control levels. By 68 weeks (16 weeks of refeeding choline), the membrane protein kinase C activity had returned to normal. At this time, 14% of the experimental animals had hepatocellular carcinoma. We suggest that choline deficiency altered the protein kinase C-mediated signal transduction within liver, and this contributed to hepatic carcinogenesis in these animals.

Supported by a grant from the American Institute for Cancer Research.

Excretion of Food-Derived Heterocyclic Amine Carcinogens into Breast Milk of Lactating Rats and Formation of DNA Adducts in the Newborn

A. GHOSHAL and E.G. SNYDERWINE

Laboratory of Experimental Carcinogenesis,
National Cancer Institute, Bethesda, MD

The distribution, DNA adduction and excretion into breast milk of 2-amino-3-methylimidazo[4,5-f]quinoline (IQ), 2-amino-3,8-dimethylimidazo[4,5-f]quinoline (MeIQx) and 2-amino-1-methyl-6-phenylimidazo[4,5-b]pyridine (PhIP) were examined in lactating female F344 rats with 5-day old pups. Six h after a single intragastric dose of radiolabeled IQ, MeIQx or PhIP to lactating dams, the radioactivity was highest in the liver and kidney followed, in descending order, by the mammary gland, omental fat, and brain. By 24 h after carcinogen administration, all tissues of the dams showed significantly reduced levels of radioactivity except for omental fat which changed only marginally from 6 to 24 h. [^{32}P]-Postlabeling analysis of mammary gland DNA indicated that the level of PhIP-DNA adducts was about five- and ten-fold higher than the levels of IQ- and MeIQx-DNA adducts, respectively. In contrast, in hepatic DNA, IQ adduct levels were approximately ten-fold higher than those from PhIP or MeIQx. The tissues of pups nursed by dams given radiolabeled IQ, MeIQx or PhIP were radioactive, indicating that these carcinogens (and/or metabolites) were excreted into breast milk. For all three compounds, the stomach contents of pups contained the highest amount of radioactivity after 6 h of suckling. The amount of PhIP-derived radioactivity was about ten-fold greater (nmole equivalents) in the stomach contents of pups than IQ or MeIQx. Radioactivity was also found in kidney, liver and urine of suckling pups. IQ-, MeIQx- and PhIP-DNA adducts were detected in the livers of pups by [^{32}P]-postlabeling. Urine from pups exposed to all three carcinogens through breast milk was mutagenic to Salmonella strain TA98 in the Ames assay in the presence of an S-9 activating system. These results indicate that breast milk is a route of exposure of the newborn to heterocyclic amines. The presence of DNA adducts in the tissues of pups further suggests that this route of exposure may have a carcinogenic consequence to the newborn.

Health Diary: A Unique Way to Empower African-Americans' Participation in Preventive Health Care

K. RAY, R. HURDLE, A. JOHNSON,
A. REDRICK, D. DEITCHER and E. WANLISS

New York State Department of Health, New York, NY

African-Americans have a 10 percent greater risk of developing cancer than whites and have a 5-year survival rate which is 30 percent lower. Cancer mortality in African-Americans is also 30 percent higher than in whites. To address these abysmal statistics, the National Cancer Institute expanded its community efforts by funding an outreach program tailored specifically for the African-American population of the United States. The program, called the National Black Leadership Initiative on Cancer (NBLIC), via the establishment of coalitions, created a nationwide network of African-American leaders who help organize, implement, and support cancer prevention programs at state and local levels.

Historically, health is not considered a priority amongst African-Americans. Dealing with other societal issues, i.e., poverty, substandard housing, unemployment, often outweighs any thoughts of preventive health. Awareness of the importance of early detection methods, and the importance of maintaining accurate personal health records and those of family members is often lacking in the African-American population. The NBLIC Manhattan Coalition sought to create a tool specifically for the African-American population that would be an attractive form of recording health appointments and health experiences. This tool, the Health Diary, encourages African-Americans to accept more responsibility for their health. The Diary is designed to enable the health care provider and the Diary user to work as a team to foster better health care. The Diary records information ranging from family medical history to a record of health care visits to tips for better living (lifestyle and food choices). Emphasis was placed an designing a tool that is culturally sensitive.

This poster describes the categories and uses of the Health Diary. Pertinent sections of the diary are presented and special attention is given to the interactive use of this document between the physician and the consumer. Recommendations for design and uses of this document are suggested.

Induction and Modulation of Mammary Preneoplastic Transformation: An *In Vitro* Model for Cancer Prevention

N.T. Telang

Division of Carcinogenesis and Prevention,
Strang-Cornell Cancer Research Laboratory,
Cornell University Medical College, New York, NY

Mammary tumorigenesis progresses via a multistep process where preneoplastic transformation represents an important intermediate step. Down-regulation of preneoplastic transformation may provide efficacious preventive intervention of tumor development. A newly developed mouse mammary epithelial cell line, derived from the murine mammary tumor virus expressing RIII strain of mouse, is utilized as a cell culture model to examine i.) expression of molecular, endocrine and cellular biomarkers for preneoplastic transformation and ii.) regulation of the perturbed biomarkers by selected retinoids, fatty acids and indoles that inhibit organ site tumorigenesis *in vivo*. The three biomarkers were expressed in response to the presence of mammary tumor virus *in vitro* prior to tumorigenesis *in vivo*. Treatment of virus expressing cell cultures individually with highest noncytotoxic doses of retinoid, omega-3 fatty acids and indole independently inhibited all the perturbed biomarkers. This *in vitro* model provides an assay to identify potential initiators and efficacious chemo- preventive agents for mammary carcinogenesis.

Mammalian Lignan Precursor in Flaxseed: Influence on Mammary Tumorigenesis

L.U. THOMPSON, M. SEIDL, L. ORCHESON and S. RICKARD

Department of Nutritional Sciences,
University of Toronto, Toronto, Ontario, Canada

The cancer protective effect of low fat, high fiber diets has been suggested to be due not only to the fiber but also to some of its associated substances. These include the mammalian lignans (enterolactone and enterodiol) with potential weak estrogenic/antiestrogenic properties that are produced by the bacteria in the human or animal colon from precursors in high fiber foods. However, because feeding purified mammalian lignan precursors has not been done, the anticarcinogenic effect of mammalian lignans or their precursors remains to be established. Thus, this study isolated a major lignan precursor (secoisolaricirecinol diglycoside; SD) from flaxseed (the richest source of precursors and previously shown to have anticarcinogenic effect when fed at 5% level), and tested its effect when provided at the promotion stage of carcinogenesis. Female rats (60; 50 days old) fed a high fat (20% corn oil) diet were gavaged with a mammary carcinogen (dimethylbenzanthracene) and one week later half of the group was given a daily gavage of 1.5 mg SD (amount equivalent to that taken in a 5% flaxseed diet) in water while the other half (control) was gavaged only with water for 20 wks. Results showed a 37% reduction ($p<0.044$) in number of tumors/tumor bearing rats or 44% reduction ($p<0.016$) in number of tumors per group with SD treatment. The tumor incidence and tumor size tended to be lower in the SD treated rats, but the difference did not reach significance. The urinary excretion of mammalian lignans in the SD treated rats was 54 times higher ($p<0.0001$) than in the control, indicating the conversion of SD to mammalian lignans. It is concluded that SD and the mammalian lignans derived from it have a protective effect at the promotion stage of tumorigenesis and that they may in part be responsible for some of the cancer protective effect of flaxseed or other high fiber foods.

Mammary Gland Expresses a Unique Pattern of C/EBP Isoforms

J.W. DEWILLE and K.S. WADDELL

*Department of Veterinary Pathobiology and
Ohio State Biochemistry Program,
Ohio State University, Columbus, OH*

Much of the controversy over the role of diet as a breast cancer risk factor stems from the lack of a biochemical mechanism linking diet and mammary tumor promotion. Our laboratory is investigating the hypothesis that diet promotes mammary tumors by altering the synthesis or function of CCAAT/enhancer binding proteins (C/EBPs). C/EBPs are DNA binding proteins that regulate the transcription of genes involved in energy metabolism, growth control, and differentiation. Three C/EBP isoforms (alpha, beta, delta) have been identified in the mouse.

Mammary tissue expresses a unique pattern of C/EBP isoforms. We provide the first evidence that lactating mammary gland and mammary adenocarcinomas express high levels of C/EBP-beta and C/EBP-delta mRNA, but little C/EBP-alpha mRNA. In contrast, liver expresses C/EBP-alpha, C/EBP-beta and C/EBP-delta mRNA. C/EBP isoform mRNA content in normal mammary tissue is not dramatically altered by proliferation, lactation, or dietary treatment. However, C/EBP-beta mRNA levels appear to be induced in high fat mammary tumors. The high level of C/EBP-beta and C/EBP-delta mRNAs in dedifferentiated mammary adenocarcinomas indicates that C/EBP expression is not confined to growth-arrested terminally differentiated cells as previously reported for liver and adipocytes.

These results suggest an important role for C/EBPs in mammary gland biology. Current experiments are aimed at defining the interaction between diet and mammary gland C/EBP isoform expression at the mRNA and protein level and assessing the potential role of C/EBPs as biochemical links between diet and mammary tumor promotion.

Mechanisms of c-*myc* Oncogene Regulation by Retinoic Acid and its Receptors

J.R. CHURCHILL,[1] M.R. ORTEGA,[1] E.O. LAM,[1] M.M. FLUBACHER[2] and R.H. CHOU[1,2]

[1]*Department of Anatomy and* [2]*Program in Molecular Biology and Biotechnology, Hahnemann University, Philadelphia, PA*

Our laboratory has shown that retinoic acid (RA) and a microinjected expression plasmid bearing a human retinoic acid receptor α (hRARα) cDNA cooperate to suppress expression from a co-microinjected human c-*myc* gene clone (pHSR-1) in *Xenopus laevis* oocytes. To test the hypothesis that a repressor domain of hRARα may be involved in the suppression of c-*myc*, we have created two hRARα deletion mutants wherein a tissue-specific regulator (hRARα-Δ300) or both a tissue-specific regulator and the repressor domain (hRARα-Δ411) are deleted. The hRARα-Δ300 construct retains the ability to suppress c-*myc* expression, but the (hRARα-Δ411) construct lacks suppressor activity. These data suggest the repressor domain of hRARα may play a role in the regulation of c-*myc* expression. We have also tested for direct interactions between the RARα-RA complex and regulatory regions of c-*myc* DNA using *in vitro* translated wild-type RARα proteins in a gel mobility shift assay. A unique gel shift band is observed with the 167bp SmaI/XhoI c-*myc* fragment bearing the P1 promoter region. No unique gel shift band was observed with DNA fragments from several other regions of the c-*myc* gene. These results suggest that RARα proteins directly interact with c-*myc* DNA.

Supported by a grant from the American Institute for Cancer Research.

Negative Effects of Chemotherapy on Nitrogen Metabolism in Healthy and Tumor Bearing Rats

T. LE BRICON, S. GUGINS and V.E. BARACOS

*Department of Animal Science, University of Alberta,
Edmonton, Alberta, Canada*

Cancer is associated with extensive wasting, but it has been suggested that cancer chemotherapy itself may participate in cachexia. Effects of chemotherapy on nitrogen (N) metabolism were investigated in Sprague-Dawley rats. Controls (n=40) and rats bearing Morris hepatoma 7777 for 7 days (n=40) were randomized to 5 treatment groups (n=8): sham, cyclophosphamide (CYP), 5-fluorouracil (5-FU), cisplatinum (CDDP) and methotrexate (MTX). Drugs were administered as a single ip injection of a high non-lethal dose. Body weight, N intake, urinary and fecal N were determined every 2 days for a period of 6 days using the Kjeldahl method; N balance was calculated for the same periods. Tumor mass of MH 7777 was 0.4 ± 0.2 g in sham treated rats. CYP had no effect on tumor mass; 5-FU induced partial remission (tumor weight <0.2 g) in ~50 % of rats. Treatment with CDDP and MTX induced total remission in more than 50% of rats. All agents studied decreased food intake to different degrees over the 6 days: CDDP (–59%) and MTX (–25%) were the most anorexogenic drugs; anorexia was significantly greater in tumor bearing rats. Chemotherapy induced progressive body weight loss and reduced N balance. Poor or negative N balances were mostly due to decreased N intake and most important changes were noted 3-4 days after drug injection.

On days 5-6 after drug injection, N balance of CDDP treated rats continued to decrease, while that of animals treated with 5-FU and MTX improved. These data suggest that negative effects of chemotherapy on N metabolism are drug specific and are more prominent in tumor bearing than in healthy rats.

Nutritional Determinants of Breast Cancer

I. ROMIEU, L. LOPEZ-CARRILLO,
E. LAZCANO and M. HERNANDEZ-AVILA

National Institute of Public Health, Mexico City, Mexico

A population based case-control study was conducted to investigate nutritional determinants of breast cancer among a Mexican population to address the following specific hypotheses: high intake of alcohol and fat increases the risk of breast cancer; high intake of vitamin A (both carotenoid and preformed vitamin A), vitamin C and E reduces the risk of breast cancer.

During the period of September 1990 to January 1993, 368 cases were diagnosed with localized breast cancers. Dietary intake was measured using a validated food frequency questionnaire designed especially for a Mexican population. Also, 1006 female population controls were interviewed.

Preliminary results are consistent with regard to the role of major risk factors such as age at menarche, age at first pregnancy, years of education, age at menopause, and family history of breast cancer. Results seem also to confirm the protective effect of longer duration of breast feeding.

In relation to food intake, the mean calorie consumption in our study population was 1902 (s.d. = 643) calories per day, the mean consumption of fat was 67.17 (s.d. = 21) g per day, and the mean consumption of saturated fat was 19.61 (s.d. = 8.8) g per day. These results suggest an increased risk of breast cancer among women who have higher caloric intake and higher fat intake that could be related to higher intake of saturated fat. Given the low alcohol consumption in our population, results are difficult to interpret; until this point, however, it seems that in postmenopausal women, alcohol intake may have an adverse effect. Smoking habits appear also to have an adverse effect on the occurrence of breast cancer. These preliminary results will be further analyzed when total sample size is completed.

Ornithine α-Ketoglutarate (OKG) Limits Muscle Protein Breakdown Without Stimulating Tumor Growth in Rats Bearing Yoshida Ascites Hepatoma

T. LE BRICON,[1] L. CYNOBER[2] AND V.E. BARACOS[1]

[1]Department of Animal Science, University of Alberta, Edmonton, Alberta, Canada; and [2]Laboratory of Biochimie A, Hopital St Antoine, 75571 Paris, France

OKG is used in the treatment of chronic malnutrition and hypercatabolic states. In cancer, OKG might similarly improve host nutritional status or stimulate tumor growth if its metabolites are limiting for tumor growth. Enteral supplementation with OKG was investigated in Sprague-Dawley rats bearing Yoshida ascites hepatoma AH130 (YAH). Tumor bearing rats were compared with *ad libitum* and pair fed controls. Rats received OKG (3.4 – 4.0 g/kg/day) or an isonitrogenous amount of glycine (GLY) (n=8 in each group). Animals were studied for 5 days and then killed for isolation of epitrochlearis muscle (EPI). OKG had no effect on weight (10 ± 1 g), N content and free amino acid levels of the tumor. In tumor bearing rats, OKG improved muscle protein balance compared to the GLY group by limiting increased catabolism due to tumor implantation.

OKG also reduced the enhanced total amino acid release from isolated EPI of tumor bearing rats by 46% compared to the GLY group. In conclusion, OKG as an enteral supplement improved muscle protein balance in rats bearing Yoshida ascites hepatoma AH130 through anticatabolic properties, without stimulating tumor growth.

Promotional Effects of Yo-Yo Dieting on Mammary Carcinogenesis

A.R. TAGLIAFERRO, A.M. RONAN,
L.D. MEEKER and H.J. THOMPSON

*University of New Hampshire, Durham, NH; and
AMC Cancer Research Center, Lakewood, CO*

Mammary cancer is the most common type of cancer among women in the U.S. Dietary fatty acid, calories and obesity have all been implicated as important risk factors in mammary tumorigenesis. It is our working hypothesis that frequent variation in circulating substrates induced by frequent bouts of caloric restriction and refeeding (i.e., yo-yo dieting) may promote mammary cancer in animals treated with a chemical carcinogen, N-methylnitrosourea (MNU). In two preliminary experiments, young female Sprague-Dawley rats were injected ip with MNU (37.5 mg/kg and 25 mg/kg respectively) at 50 days of age and were fed individually a purified AIN-76A diet modified in fat (corn oil 24% w/w). Dieting and pair fed control groups were fed in two three-hour meals daily (ME-D and ME-C respectively) and a second control group was fed *ad libitum* (Ad lib-C). ME-D were restricted to consume 67% of the Ad lib-C intake for 1 week followed by three weeks refeeding over 4, four-week cycles. ME-D and ME-C had similar body weights and body fat mass weights; both were significantly less than Ad lib-C. However, rate of mammary tumor appearance in the ME-D group was significantly higher than both ME-C and Ad lib-C. In a second experiment, the incidence of tumorigenesis among ME-D group was 20% and 13% greater than ME-C and Ad lib-C respectively. There were no differences in tumor number or tumor weight among treatment groups. Weekly measurement of energy and substrate utilization by indirect calorimetry indicated there were no differences in energy expenditure among the three groups, but there were differences in substrate utilization. ME-D utilized more fatty acids and less glucose during dieting and more glucose and less fatty acids during weight recovery, than the control groups. Present findings support our hypothesis and suggest cancer promotion may be related to a shift in glucose and fatty acid utilization induced by yo-yo dieting.

Supported by the American Institute for Cancer Research grant 91A21.

Reduced Hypothalamic Release of Neuropeptide Y into Microdialysates of Anorectic Tumor-Bearing Rats

W.T. CHANCE, A. BALASUBRAMANIAM and J.E. FISCHER

*Department of Surgery, University of Cincinnati
Medical Center and Veterans Affairs Medical Center,
Cincinnati, OH*

The potent orexigenic effect of neuropeptide Y (NPY) caused us to hypothesize that NPY feeding systems might be altered in anorectic tumor-bearing (TB) rats. In previous work we demonstrated that NPY-induced feeding was attenuated and NPY levels were decreased in the hypothalamus of TB rats. To determine whether these alterations were associated with decreased release of NPY, 20 gauge microdialysis guide cannulae (BAS, W. Lafayette, IN) were implanted 2 mm above the perifornical hypothalamus of 21 male Fischer 344 rats. Two weeks later six of these rats were inoculated in the dorsum with methylcholanthrene sarcoma, while the remaining rats formed *ad libitum*-fed freely-feeding (FF) and food-restricted matched carcass weight (MCW) control groups. Body weight, intake of rat chow and water consumption were monitored daily. Microdialysis of the hypothalamus was conducted in all rats using a 2 mm loop (BAS) and flow rate of 2µl/min (artificial CSF) prior to (day 18) and after (day 27) the development of anorexia. Concentrations of immunoreactive NPY were determined in 100µl of dialysates by RIA. Although there was no difference between groups in immunoreactive NPY in dialysates collected prior to the onset of anorexia, NPY levels in post-anorexia dialysates of TB rats (1.8 + 1.1 pmol/ml) were decreased significantly compared to either FF (28.9 + 9.8 pmol/ml) or MCW (18.1 + 5.7 pmol/ml) control groups. The absence of significant alterations in the MCW control group indicates that the decrease in immunoreactive NPY in anorectic TB rats was not secondary to nutritional factors. These results suggest that NPY feeding systems are altered in TB rats and that reduced release of hypothalamic NPY may be directly involved in the etiology of experimental cancer anorexia.

Time Dependence of the Relationship Between Diet and Breast Cancer in 34 Countries

B.P. DUNN

British Columbia Cancer Agency,
Vancouver, British Columbia, Canada

It is not known when in life diet acts to influence the risk of breast cancer. Many countries have had substantial changes in diet in the last several decades, offering the opportunity to investigate this question in an international correlation study. For 34 countries, total fat and grain consumption from 22 published estimates of diet covering a period of 25 years were used to generate smoothed estimates of the consumption of these food items for 5 year intervals from 1960 to 1985. These values were compared with 1985 breast cancer incidence data for 5 year age intervals in order to investigate the effect of lag period on the correlation between diet and cancer risk.

Fat consumption was positively associated and grain consumption was negatively associated with breast cancer in all age categories. Fat and grain consumption were approximately equally predictive of premenopausal breast cancer, with optimum lag periods of 15 to 20 years and only a modest time dependence. For postmenopausal women, grain consumption predicted cancer better than fat consumption, and was optimally measured 20 years prior to cancer. For postmenopausal women, the predictive value of fat was strongly time dependent, and declined sharply as the lag period was shortened. Highest correlations were obtained for fat consumption evaluated at the beginning of the 25 year span of the diet data. Extrapolation suggests that the optimal predictive value of fat consumption for the risk of postmenopausal cancer would occur if consumption had been evaluated in premenopausal years, with lag periods of up to 40 years. In comparative studies, the correlation between fat and grain consumption and female colon cancer showed only a modest time dependence, and an optimum lag period of approximately 20 years for fat, and 0 to 5 years for grain.

The data indicate a lag period of 2 decades or more in the effect of diet on breast cancer risk. In populations where individuals change their diet with time, diet assessment near the time of cancer diagnosis (case control studies, and cohort studies with short followup) may not be optimal for demonstrating an association between diet and breast cancer.

Addition of Vitamin B_{12} and Folate to a Severely Methyl-Deficient Diet Does Not Alter the Diet's Capacity for Causing Rapid and Persistent Changes in DNA Methylation in Rat Liver

J.K. CHRISTMAN, M.L. CHEN, G. SHEIKHNEJAD, S. ABILEAH and E. WAINFAN

Molecular Biology Program, Michigan Cancer Foundation and Molecular Oncology Program, Meyer L. Prentis Comprehensive Cancer Center of Metropolitan Detroit, Detroit, MI

It has been amply demonstrated that dietary lipotrope deficiency can promote and/or cause development of liver tumors in rodents. We have shown that a severely methyl deficient diet [MDD-amino acid defined minus choline, methionine, vitamin B_{12} and folate and supplemented with homocystine] for 1-4 weeks leads to a decrease in the overall level of DNA and tRNA methylation, an increase in the activity of DNA and tRNA methyltransferases and alterations in the patterns of methylation and levels of transcripts for several growth related genes in the livers of rats. Only the alterations in pattern of methylation of specific genes persist when an adequate level of lipotropes is restored to the diet.

Both the rapidity with which changes in DNA methylation occur and their persistence for > 12 wks in the absence of further methyl deprivation, suggest that changes in DNA methylation may play a contributory role in causing lipotrope-deficiency induced liver tumor development. However, rats do not survive long enough on MDD to develop tumors. To determine if feeding of MDD has more profound early effects on DNA methylation in the liver than a diet which allows the animals to survive to the point where tumors develop, the effects of MDD were compared with those of a diet deficient only in choline and methionine [MDDV-MDD supplemented with B_{12} and folate]. Within 4 days, both diets induced similar alterations in liver cell morphology, lipid accumulation and mitogenic index. Liver DNA from animals fed MDDV became hypomethylated at the same rate and to approximately the same extent as liver DNA from animals fed MDD.

Restoration of lipotropes to the diet of MDD- or MDDV-fed rats decreased the mitotic index to normal within days. No differences between the two groups could be detected in the rate or extent of restoration of normal liver morphology or lipid content. Most importantly, loss of methylation at specific sites in several growth regulatory genes was equally resistant to reversal in both the MDD- and the MDDV-fed rats. These results demonstrate that B_{12} and folate do not moderate the immediate effects of choline/methionine deficiency on rat liver even though they have a major influence on the overall health of the animals.

Supported by the American Institute for Cancer Research Grant 92A35 and the Lloyd and Marilyn Smith Fund.

Exposure to Dietary Fat *in Utero* Affects Subsequent Development of DMBA-Induced Mammary Tumors

L.A. HILAKIVI-CLARKE, D. MURPHY and M.E. LIPPMAN

*Lombardi Cancer Research Center,
Georgetown University, Washington, DC*

We investigated whether dietary manipulations *in utero* influence DMBA-induced mammary tumorigenesis in Sprague-Dawley rats. Female animals were fed either with a high-fat diet (n-6 polyunsaturated fatty acids, containing 46% of calories from fat) (n=9); a low-fat diet (12% fat) (n=9); or a control laboratory diet (20% fat) (n=10) throughout pregnancy. The food consumption and body weight gain of the pregnant rats, and the number of pups born and their body weights in different groups did not differ. The female offspring, which were given a control laboratory diet from postnatal day 1 onwards, were treated with 10 mg of DMBA at the age of 55 days. The proportion of animals with mammary tumors and the number of tumors per group were significantly higher among the rats which were given high-fat diet prenatally (n=40) than in the low-fat (n=40) or control-fat diet (n=36) groups (Gehans-Wilcoxan test; p <.03 and p <.05). Lowest tumor incidence was noted in rats fed with low fat diet *in utero*. The size of the mammary tumors upon first detection or the growth rate of aggravating tumors did not differ between the three groups. To address the biological mechanisms through which the association between early dietary manipulations and subsequent mammary tumor growth might have been mediated, we measured total plasma levels of 17β-estradiol (E2) in the plasma. The plasma levels of total E2 were significantly higher in those pregnant females which were fed with high-fat diet than in the two other dietary groups (p <.05) (F(2,14)=4.91, p <.02).

The results suggest that exposure to high dietary fat *in utero* induces a permanent increase in the growth of DMBA-induced mammary tumors in female rats. A low fat diet appears to reduce mammary tumorigenesis. Elevated levels of plasma E2 *in utero* observed in the high-fat diet group may mediate this effect.

Weight at Age 30 and Breast Cancer Risk

N.B. KUMAR, N.M. AZIZ, D.V. SCHAPIRA,
C.P. COX and G.H. LYMAN

*H. Lee Moffitt Cancer Center and
Research Institute and University of South Florida
College of Medicine and Public Health, Tampa, FL*

Obesity has been associated with the incidence of breast, endometrial and ovarian cancers. Significant changes in weight and body fat distribution can occur in women from puberty, through childbearing years and beyond. The objective of the study was to observe if there was any pattern to the weight gain between age 16 to adulthood and beyond in breast cancer patients and controls, and consequently to assess if there was a significant decade in a woman's life contributing to breast cancer risk. Anthropometric, medical and hormonal histories were obtained from 218 consecutively recruited, newly diagnosed breast cancer patients admitted to the H. Lee Moffitt Cancer Center and Research Institute, and 436 community-based controls, matched in a 2:1 ratio for age and menopausal status. A weight gain of 15 pounds or over was observed in 63.8% of cases as compared to 49.3% controls (P = 0.0006) from age 30 to current age. Similarly, over 48% of cases gained over 15 pounds from ages 16 to 30 as compared to 37% (P = 0.01) of the control population. Although weight gain from age 16 onwards was significantly higher in breast cancer patients at each decade when compared to controls, a particularly significant and independent association between weight at age 30 (P = <0.0001) and risk of breast cancer was noted.

Women who progressively gain weight from puberty to age 30 and specifically in the third decade of life should be considered at a higher risk for developing breast cancer.

Tumor Prostaglandin E_2 Levels and Receptor Tyrosine Kinases May be Related to Tumor Growth via Dietary Fat

R.C. REITZ, D.T. COLOMBO and M. NUTTER

*Department of Biochemistry,
University of Nevada, Reno, NV*

Dietary fat plays an important nutritional role in the etiology of human cancer in general and breast cancer specifically. The $\omega 6$ fatty acids promote tumor growth, and the $\omega 3$ fatty acids inhibit tumor growth. We have used diets containing 4 different oils in an attempt to focus on the mechanisms by which the polyunsaturated fatty acids exert control over tumor growth. These fats were fed at 10% (W/W). Corn oil served as the control, and the other three oils were rich in either eicosapentaenoic acid (EPA), docosahexaenoic acid (DHA), or a mixture of both EPA and DHA. Tumor growth was inhibited the most by the mixture of EPA and DHA, while the diet rich in EPA produced the smaller effect on tumor growth. The diet rich in DHA resulted in a growth reduction intermediate between EPA and the mixture. The fatty acid composition of the tumors tended to reflect the dietary fatty acids, and the data did not clearly suggest a particular $\omega 3$ fatty acid which should be emphasized. We found significant reductions in the concentration of prostaglandin E_2 in the tumors from the animals fed the $\omega 3$ diets, again without any real specificity relating to a particular dietary $\omega 3$ fatty acid. We then measured the activity of the receptor tyrosine kinases (RTK) using insulin growth factor (IGF) and epidermal growth factor (EGF) as the stimulating ligands. IGF resulted in the greatest stimulation of activity in the corn oil-fed control animals. The different $\omega 3$ dietary fats produced a decrease in IGF stimulation of RTK which mimicked the decreased tumor growth. EGF-stimulated RTK activity also decreased in a pattern similar to tumor growth inhibition. These data strongly implicate RTK in the mechanism by which the $\omega 3$ fatty acids produce a reduction in tumor growth.

Supported in part by the Reno Cancer Foundation.

Epigenetics of Neoplastic Transformation: Essential Background for Nutritional Studies

H. RUBIN, A. YAO and M. CHOW

*Department of Molecular and Cell Biology and Virus Laboratory,
University of California, Berkeley, CA*

The conventional view about the origin of cancer shifts from genetic to epigenetic depending on the currently dominant fashion of thought in biology. With the current supremacy of molecular genetics, certitude is commonly expressed that somatic mutation is the primary driving force in cancer. Given this point of view it hardly seems likely that diet will influence the course of cancer. The current genetic point of view received its strongest impetus from finding diverse mutations in human tumors, mostly in their later stages. But this merely reveals association not causation. Indeed it is difficult if not impossible to establish causality by examining the product of a process extended over a lifetime in organisms in which one cannot do a thorough, quantitative analysis. The answers seemed to lie in the study of cell transformation in culture. Studies by Heidelberger, Kennedy and others in the 1970s and early 1980s showed that exposure of cells to chemical or physical carcinogens increased the probability that **all** the treated cells, or their descendants, would undergo neoplastic transformation. This result is strongly at variance with the low frequency expected from a mutational origin of the transformation. These observations have largely been forgotten in the rising tide of molecular genetics and its tumor-related extensions. But another reason for the lapse is that the results in cell culture were erratic, varying quantitatively from experiment to experiment, and they generally gave low efficiencies of transformation on a per cell basis.

Over the past five years we have been developing and utilizing a cell culture system of transformation with a much higher efficiency of "spontaneous" transformation, and have been learning how to limit its variation as well as direct its capacity for transformation. The results, which will be outlined, show that "spontaneous" transformation is not spontaneous, but depends very much on the physiological state of the cells, in particular the growth constraints of crowding and low concentrations of serum. It is, to all intents and purposes, an epigenetic and adaptive process, rather than a genetic process, although some role of mutations in the long term progression of tumors in humans is not ruled out.

The establishment of an epigenetic foundation for neoplasia signifies an important role for nutrition in modifying the susceptibility of cells to transformation and even in reversing the transformed state. We will describe the recent improvements in controlling the capacity of cells for transformation which is a prerequisite to using them intelligently for nutritional studies.

Comparative Study of Effects of Caloric Restriction on the Metabolic Activation of Aflatoxin B_1 and Benzo[a]pyrene

K.T. CHUNG,[1] W. CHEN,[2] Y.G. ZHOU,[2] J. NICHOLS,[2] P.P. FU,[2] R.W. HART[2] and M.W. CHOU[2]

[1]Department of Biology, Memphis State University, Memphis, TN; and [2]National Center for Toxicological Research, Jefferson, AR

Caloric restriction (CR) lowered chemically-induced tumor incidences in laboratory animals. Previous results indicated that the metabolic activation of aflatoxin B_1 (AFB_1) can be inhibited by CR. In this presentation, a comparison of the effect of CR on the *in vivo* and *in vitro* metabolic activation of AFB_1 and benzo[a]pyrene (BaP) in terms of the carcinogen-DNA adduct formation was studied. In male F344 rats, CR increased hepatic microsomal aryl hydrocarbon hydroxylase and cytochrome P-450IA1 dependent ethoxy-resorufin O-deethylase activities significantly. The *in vitro* microsome-mediated binding of BaP to calf thymus DNA was also enhanced by 45%. However, the hepatic microsomal AFB_1 metabolizing enzyme activity and subsequent microsome-mediated AFB_1-DNA binding were significantly decreased by CR. Similar results obtained from the *in vivo* study revealed that CR increased the total BaP-DNA adduct formation in liver of rats pretreated with [³H]BaP, and decreased the formation of AFB_1-DNA adduct pretreated with [³H]AFB_1. Our results indicate that the effect of CR on metabolic activation of xenobiotics is dependent upon the nature of the chemical carcinogens and selected xenobiotic metabolizing enzymes whose activities may be significantly altered by CR.

Dietary Effects on Gene Expression in Mammary Tumorigenesis

P. ETKIND

*Montefiore Medical Center, Albert Einstein
College of Medicine, Bronx, NY*

Current work in our laboratory strongly suggests that expression of specific genes involved in mammary tumorigenesis can be affected by a high fat diet. We have carried out these studies in the C3Hf mouse in which the events of spontaneous mammary tumorigenesis are clearly understood on a molecular/genetic level. In the C3Hf mouse, offspring receive no exogenous mouse mammary tumor virus (MMTV), the known etiologic agent of mammary adenocarcinoma in most inbred strains of mice, but rather develop mammary tumors due to the transcriptional activation, after a number of pregnancies, of one specific endogenous MMTV proviral DNA sequence present at the Mtv-1 locus.

Mammary glands of multiparous C3Hf mice produce the MMTV specific RNA transcripts of 9.0 and 3.8 kilobases (kb) which result from transcription of the Mtv-1 locus, but these transcripts are never seen in the mammary glands of C3Hf mice until second parity (pregnancy). C3Hf mice who do not undergo pregnancy do not produce these MMTV specific transcripts and do not develop breast tumors.

The lactating mammary glands of C3Hf mice which had been fed either a low fat (5% corn oil) or high fat (23.5% corn oil) diet upon weaning (21 days) were analyzed for the presence of MMTV transcripts derived from the Mtv-1 locus. We have detected the presence of the 9.0 and 3.8 kb Mtv-1 specific transcripts in the lactating mammary glands of 6 out of 7 **first** parity C3Hf mice on a high fat diet. None of the fifteen first parity C3Hf mice on a low fat diet which we have analyzed have shown expression of these MMTV specific transcripts in their lactating mammary glands. In addition at this time 4 out of 20 C3Hf mice which have been maintained on a high fat diet have developed mammary tumors at the age of 11-12 months (3 after 3 pregnancies, 1 after only 1 pregnancy), suggesting that the high fat diet is decreasing the normal latency period (18-20 months) for tumor development.

Supported by the American Institute for Cancer Research grant.

The Effect of Diet, Ethnicity and Total Body Fat on Non-Protein Bound Estradiol Levels

L.A. JONES, W.J. KLISH, R. SHALLENBERGER,
D. JOHNSTON, M. FOLLEN-MICHELL, V. VOGEL,
V.B. GOSSIE, J. YICK, C. HARRIS,
C. GALVEZ and W. INSULL

*Experimental Gynecology/Endocrinology, University of Texas
M.D. Anderson Cancer Center, Houston, TX; Lipid Research
Clinic and the Departments of Pediatrics and Medicine,
Baylor College of Medicine, Houston, TX*

Breast cancer constitutes 26% of all cancers that occur in women, while eighteen percent of all female cancer deaths can be attributed to breast cancer. Epidemiological studies have implicated ovarian hormones in the etiology of breast cancer. However, no unifying endocrine mechanism explaining the role of estrogen has yet emerged. It is also well-known that the incidence of breast cancer varies according to ethnicity. Although several studies have indicated that diet may play a role in the etiology of breast cancer, no unifying mechanism has emerged. Therefore, the purpose of this study is: (1) to determine if in women who reduce their dietary fat from the usual American level of 40% to 20% of total calories there is a change in their blood hormone levels and (2) to determine what role, if any, ethnicity has. At present 40 postmenopausal women, 13 white, 16 African-Americans, and 11 Hispanic-Americans, have been randomized according to % total body fat to either a control or intervention group. All women were disease free, had no personal history of breast cancer, and were not already on a low fat diet. Fasting blood was collected in heparinized tubes and refrigerated within 30 minutes after blood drawing. After refrigeration overnight, plasma samples were aliquoted into 2 ml vials and stored at −70°C until analyzed. Blood samples were analyzed using the ultracentrifugation isodialysis assay. Percentage total body fat was determined utilizing Total Body Electrical Conductivity.

From our preliminary results, it seems that % of total body fat has a different effect on the level of % of nonprotein bound estradiol found in the three ethnic groups: African-Americans 40.9/1.83; whites 40.4/1.73, and Hispanics 40.9/2.05. Additional data regarding our findings on these ethnic groups will be presented.

Effect of Eicosapentaenoic Acid on the Incorporation of Linoleic Acid in a Human Breast Cancer Cell Line

M.A. HATALA and D.P. ROSE

American Health Foundation, Valhalla, NY

Linoleic acid (LA), an n-6 fatty acid, stimulates the growth of a human breast cancer cell line, MDA-MB-435, when cells are grown in serum-free media. The mechanism for this effect may involve the metabolism of LA to arachidonic acid, and the utilization of phospholipid-incorporated arachidonate for eicosanoid biosynthesis. In the present study, the dynamics of the cell uptake and the incorporation of LA into major lipid pools were defined for MDA-MB-435 cultured cells. In addition, the effect of exogenous eicosapentaenoic acid (EPA), an n-3 fatty acid, on LA cell uptake and lipid distribution were investigated. Cells were plated at 2.0×10^4 cells/cm^2 and exposed to the fatty acids for 24 hours in serum-free media. Approximately 50% of [^{14}C]LA (1.28 µg/ml; 1.0 µCi) was taken up by the cells by 8 hours and ~90% by 24 hours. Eighty percent of the incorporated radioactivity was found in the phospholipid fraction with the remaining 20% in neutral lipids. Analysis of the relative distribution of [^{14}C]LA among the phospholipid classes revealed that ~65% of phospholipid-associated LA was found in the phosphatidylcholine, 21% in the phosphatidylethanolamine, and 14% in the phosphatidylinositol/serine fractions. Although the uptake of [^{14}C]LA was not attenuated by the addition of EPA to the cells at 1.0, 4.0, and 16.0 µg/ml, in the presence of EPA there was a concentration-dependent shift of [^{14}C]LA distribution from the phospholipid to the neutral lipid compartment. However, this shift did not alter the relative distribution of the three phospholipids within the phospholipid compartment. EPA dose-dependently decreased [^{14}C]LA incorporation into each of the three classes of phospholipids; at the highest EPA dose (16 µg/ml) incorporation of [^{14}C]LA was 43.1, 38.3, and 45.2% of control for phosphatidylcholine, phosphatidylethanolamine, and phosphatidylinositol/serine, respectively. These results suggest that EPA reduces incorporation of LA into cell membrane phospholipids and promotes redistribution of LA to the neutral lipid compartment. This provides one mechanism by which n-3 fatty acids may inhibit MDA-MB-435 cell growth.

Evidence for the Involvement of Exogenous Lipids in Cell Proliferation Regulated by *ras* Protein

P. HOMAYOUN

Department of Molecular Biology,
Cleveland Clinic Foundation, Cleveland, OH

Dietary fat has been reported to enhance the growth of both spontaneously and chemically induced colon and mammary tumors. However, the exact mechanism by which dietary lipids exert their effects remains to be determined. The Ha-*ras* proto-oncogene product, p21^{ras} is involved in cell proliferation. Mutants of p21^{ras} are frequently observed in a number of tumors. p21^{ras} is active when bound to GTP and becomes inactive upon GTP hydrolysis. This process is induced by cytoplasmic GTPase activating protein (GAP). GAP activity has been reported to be inhibited by polyunsaturated fatty acids, particularly arachidonic acid. In addition, GAP inhibitory lipids were observed after serum stimulation of quiescent cells. This investigation has been performed to determine the physiological significance of this inhibition and to identify the origin of these lipids. BALB/c3T3 cells were rendered quiescent and treated as follows: First, cells were stimulated by 10% fetal calf serum (FCS). Cellular lipids were then extracted and separated by thin-layer chromatography (TLC). Lipids from different fractions were incubated with GAP and [α-^{32}P]GTP-loaded *ras* protein. The hydrolysis of p21^{ras}-GTP was determined by immunoprecipitation with anti-*ras* antibody followed by separation of nucleotides by TLC. In the second series the cells were labelled with [^{3}H]arachidonic acid prior to serum stimulation. The percentage of radioactivity was measured after extraction and separation of lipid fractions. In the third series, quiescent cells prelabelled with [^{3}H]arachidonic acid were stimulated by media containing [^{14}C]linoleic acid. The percentages of different labels were determined after extraction and separation of cellular lipids.

Results have shown that: 1. Serum stimulation of quiescent cells resulted in the production of lipids with the ability to inhibit GAP activity. 2. Labelling of cells with [^{3}H]arachidonic acid indicated that inhibitory lipids did not seem to be arachidonic acid. 3. Stimulation of cells with media containing [^{14}C]linoleic acid demonstrated that [^{14}C]label had been taken up by cells and appeared at the level of previously observed GAP inhibitory lipids.

These results suggest that extracellular lipids (fatty acids) transported by cells might be involved in the stimulation of cell proliferation regulated by *ras* protein.

Growth Inhibition of MCF-7 Human Mammary Cancer Cells by Inositol Hexaphosphate (InsP$_6$)

I. VUCENIK,[1] G. YANG,[2] A. MEYER[2] and A.M. SHAMSUDDIN[2]

[1]*Department of Medical and Research Technology; and*
[2]*Department of Pathology, University of Maryland*
School of Medicine, Baltimore, MD

Myo-inositol hexaphosphate (InsP$_6$ or phytic acid) is a ubiquitous plant component that constitutes 1-5% by weight of most cereals, nuts, legumes, and oil seeds.

Recent studies have shown that InsP$_6$ is chemopreventive and chemotherapeutic in both *in vitro* and *in vivo* models, viz. rat and mouse colon carcinoma, transplanted and metastatic fibrosarcoma (rat and mouse), rat hepatoma, also human colon cancer (HT-29) and K562 erythroleukemia cells. Most recently we and others have demonstrated antitumor and antiproliferative action on rat and mouse mammary cancer.

In the present study we have investigated the effect of InsP$_6$ on human mammary cancer cell line MCF-7 on: a) growth inhibition (IC$_{50}$-MTT assay); b) cell morphology; c) uptake; d) intracellular distribution (differential centrifugation); e) metabolism (anion exchange chromatography).

A 50% inhibition of cell growth (IC$_{50}$) was found with $\geq$1.0 mM InsP$_6$. InsP$_6$ treated cells also became smaller. As early as 1 min after incubation of MCF-7 cells with [^{3}H]-InsP$_6$ (SA 444 GBq/mmol, 370 Bq/10^6 cells) 3.1 $\pm$ 0.7% of [^{3}H]-InsP$_6$ was taken up by MCF-7 cells, and 9.5 $\pm$ 1.6% after 1 hr. By differential centrifugation 86% radioactivity was recovered from the cell cytosol. Ion exchange chromatography showed that 58% of the absorbed intracellular radioactivity was in InsP$_6$ form.

Our data suggest that InsP$_6$ is rapidly taken up by MCF-7 cells, and it is an effective inhibitor of mammary cancer cell growth *in vitro*.

Supported by the American Institute for Cancer Research grant 92B18-REV(I.V.).

In Vitro Treatment of Human Breast Cancer Cells with RRR-α-Tocopheryl Succinate (Vitamin E Succinate) Inhibits Proliferation and Enhances the Secretion of Biologically Active Transforming Growth Factor-Beta (TGF-β)

B. ZHAO, M. SIMMONS-MENCHACA, A. CHARPENTIER, K. ISRAEL, B. SANDERS and K. KLINE

Division of Nutritional Sciences and Department of Zoology, University of Texas at Austin, Austin, TX

RRR-α-tocopheryl succinate (vitamin E succinate, VES) at 15, 10, 5, and 1 µg/ml inhibits the *in vitro* proliferation of human breast cancer cells. Treatment of MCF-7-BK, MCF-7-McGuire, MDA-MB-231, and MDA-MB-435, breast cancer cells for 24 or 48 hours with 10 µg/ml of VES inhibited cell proliferation by 41, 41, 52, and 45%, respectively at 24 hours; and by 58, 75, 81, and 71%, respectively at 48 hours. Cell conditioned media from the four breast cancer lines, treated for 24 hours with VES, exhibited antiproliferative activity that ranged from 31 to 48% when tested in the Mv1Lu-CCL-64 mink lung bioassay cells. Heat treatment (100°C for 3 minutes) enhanced the antiproliferative activity of the conditioned media, creating a 54-66% inhibition of mink lung cell proliferation. Bio-Gel P-60 gel filtration column chromatography of cell conditioned media from VES treated breast cancer cells revealed two acid stable antiproliferative factors of M_r 44,000 and 14,000. Antibody neutralization studies of the M_r 14,000 antiproliferative factor with 2.5, 1.25, 0.6, and 0.3 µl of a pan-TGF-β-specific reagent that neutralizes the activity of all three mammalian TGF-β isoforms (TGF-β 1, 2, and 3) enhanced the proliferation of the Mv1Lu-CCL-64 mink lung bioassay cells by 310, 268, 231, and 120 percent, respectively. Northern blotting analyses of total RNA and poly A+RNA from VES-treated MCF-7 and MDA-MB-435 breast cancer cells showed elevated levels of TGF-β3 mRNA. Two types of analyses suggest that VES inhibits breast cancer proliferation via some mechanism in addition to induction of biologically active TGF-β. Namely, pan-specific antibodies specific for TGF-βs are not capable of completely neutralizing the antiproliferative activity of the unfractionated cell conditioned media from VES-treated breast cancer cells, and the growth inhibition profile of breast cancer cells treated with VES is different from the growth inhibition profile obtained when the cells are treated with purified TGF-β1 or 2, singly or in combination. These data suggest that VES inhibits human breast cancer proliferation by inducing biologically active TGF-β, as well as by some additional mechanism(s).

Supported by National Cancer Institute Grant CA-45422 from the National Institutes of Health, Bethesda, MD and the Foundation for Research, Carson City, NV.

Influence of Dietary Linoleic and Stearic Acids on Phosphatidylinositol Metabolism in Transplanted Murine Mammary Tumors

A.S. BENNETT, L.A. SHAFFER, J. SMITH-HALL,
D.E. PEAVY AND R.L. YOUNG

Ball State University, Muncie, IN; and
Indiana University School of Medicine; and
Veterans Affairs Medical Center, Indianapolis, IN

Although the mechanisms involved are not clear, numerous studies have demonstrated the tumor promoting effects of fatty acids in the murine mammary system. Certain polyunsaturated fatty acids, such as linoleic acid, are required for the growth of transplanted tumors and increase metastases by influencing the lodgment, implantation and survival of mouse mammary cells. Stearic acid has been shown to inhibit growth of normal and neoplastic mammary epithelial cells, to inhibit tumor formation in rats and to be the most cytotoxic fatty acid in inhibiting colony-forming abilities of five human cancer cell lines. Research suggests that interactions with the metabolism of phosphatidylinositol (PI) and phosphotyrosine-containing proteins may be involved in tumor growth and development.

Murine mammary tumors transplanted into female Strain A/ST mice fed either 5% fat stock diets (ST) or 15% fat diets, rich in either 18:2 (LA) or 18:0 (SA), were used to (1) assay for phospholipase C (PLC) activity using tritiated phosphatidylinositol as a substrate; (2) detect phosphotyrosine-containing proteins by Western blotting using [125]-I-antiphosphotyrosine monoclonal antibodies; and (3) identify histological characteristics by light microscopy. Tumors were excised when they reached 0.2-3.0 g. Small tumors (less than 300 mg) from mice fed LA diets, but not those fed SA diets, had elevated PLC activity. Tumors from LA fed mice had enhanced tyrosine phosphorylation patterns. Changes in morphology were observed as the tumors grew larger. Results indicate an influence of dietary LA and SA on phosphatidylinositol metabolism and mammary tumorigenesis in this model.

Supported in part by Veterans Affairs General Medical Research funds and grants from Delaware County Cancer Society and Indiana Academy of Science.

On the Mechanism of the Dietary Regulation of Breast Cancer

N.H. SARKAR, H.W. LI and W. ZHAO

Medical College of Georgia, Augusta, GA

The mechanism by which calorie restricted diet suppresses mammary tumor development in experimental animals is poorly understood. Mammary tumors in mice result from transcriptional activation of *int* oncogenes caused by mouse mammary tumor virus (MMTV). In order to determine how a calorie restricted diet affects the process of mammary tumorigenesis, we fed RIII mice with low calorie (LC; 10 kcal/day) and high calorie (HC; 16 kcal/day) diets. The milk-borne MMTV that RIII mice carry is biologically distinct from the virus carried by C3H mice, a strain that has primarily been used in previous dietary studies. Our results, as observed by Northern and slot blot hybridizations, show that mice fed a LC diet for various periods of time express 4-16 fold less MMTV-RNA than those mice fed a HC diet. In addition, there was, as determined by radioimmunoassays, a 1.7-2.4 fold decrease in the levels of circulating prolactin (Prl) in LC diet fed mice as compared to HC diet fed mice. The tumor incidence was 80% less in the former group of mice than in the latter group. In contrast, there was no effect of calorie restricted diet on the levels of pituitary Prl-RNA.

Taking into account the fact that steroid hormones, such as progesterone (PR), regulate MMTV biosynthesis, we envision the following pathway through which LC diet may suppress mammary tumor development. LC diet, by lowering serum Prl, may lower the level of PR which, in turn, lowers the expression of MMTV via an interaction that involves the binding of PR receptors to the long terminal repeat (LTR) of MMTV. This reduction in MMTV expression results in a lower probability of an MMTV provirus being integrated near and activating an *int* oncogene, and thereby reduces the frequency of mammary cell transformation and subsequent tumor induction. Possibility also remains that dietary calorie may directly alter the level of other hormones, such as estrogen, PR and/or their receptors, that result in regulating in a complex manner the expression of MMTV. Diet may also affect the growth of mammary tumor cells by affecting the expression of certain growth hormones. In any event, since LC diet does not seem to affect Prl-RNA, an important question that needs to be addressed now is how a LC diet lowers the level of serum prolactin, and to determine whether or not a LC diet may directly lower the expression of oncogenes involved in mammary cell transformation.

Supported by grants from the National Institutes of Health and the American Institute for Cancer Research.

Palmitate Inhibits Epidermal Growth Factor Stimulated DNA Synthesis Possibly Via Epidermal Growth Factor Receptor Associated G-Protein

R.W. HARDY, H. JO, A. WELLS,
J. WILLIFORD and J.M. MCDONALD

*Department of Pathology, University of Alabama at
Birmingham, Birmingham, AL*

The role of individual fatty acids in promoting or reducing cancer risk remains unclear. We have found that in Hs68 human fibroblasts, as little as 100 µM palmitate (C:16-0) inhibits epidermal growth factor (EGF) stimulated DNA synthesis (as measured by ^{3}H thymidine uptake). Palmitate inhibition of EGF stimulated DNA synthesis was concentration dependent and had no effect on epidermal growth factor receptor (EGFR) tyrosine kinase activity. We have found that an ADP-ribosylated 41 kDa G protein is specifically co-immunoprecipitated from NR6WT (mouse fibroblast cells expressing human EGFR). The addition of EGF increased the amount of ribosylated 41 kDa G-protein that co-immunoprecipitated with the EGFR. Using an anti-G_i antibody we find that EGF increases the total mass of G-protein co-immunoprecipitated with the EGFR. Palmitate decreases the amount of ribosylated 41 kDa G-protein that co-immunoprecipitates with the EGFR, while myristate had no effect. The influence of palmitate on the EGFR associated G-protein may be the underlying mechanism whereby palmitate inhibits cell growth.

Vitamin E Succinate Inhibits Protein Kinase C Activity: Correlation With its Inhibitory Effects on Growth and Invasion of Breast Carcinoma Cells

R. GOPALAKRISHNA, U. GUNDIMEDA, and Z.-H. CHEN

Department of Pharmacology and Nutrition,
School of Medicine, University of Southern California,
Los Angeles, CA

Both the succinate and phosphate esters of vitamin E (*d*-α-tocopherol) at low concentrations (1 to 25 µM) inhibited both phosphotransferase activity and phorbol ester binding of purified PKC. However, vitamin E, its charge-free esters acetate and nicotinate, and the water soluble analog Trolox all failed to inhibit PKC even at a high (250 µM) concentration under these conditions. An increase in ionic strength enhanced the inhibition of PKC activity induced by succinate and phosphate esters of vitamin E and decreased the succinate ester-induced inhibition of PDBu binding. Among the various analogs of vitamin E tested, only vitamin E succinate inhibited the growth of various human breast carcinoma cells in culture. Both estrogen receptor positive and negative cell lines were affected by vitamin E succinate. Vitamin E succinate also induced a programmed cell death (apoptosis) in these cell lines. Invasive potentials of various breast carcinoma cell lines were evaluated by using micropore filters coated with reconstituted basement membrane (matrigel). The invasion of these tumor cells was also decreased by vitamin E succinate but not by other analogs of vitamin E.

This is the first report on vitamin E succinate inhibition of an enzyme which is related to signal transduction involved in growth and transformation. Vitamin E phosphate, although it was very effective in inhibiting PKC activity in a test tube, did not inhibit either growth or invasion of all these cell lines probably due to its rapid hydrolysis by intracellular phosphatases. This further suggests that increased release of vitamin E inside the cell alone cannot induce inhibitory affects on growth and invasion, but intact charged ester is also required. Although unhydrolyzed vitamin E succinate may have some unique effects on the cell, the hydrolytic product vitamin E, by acting as an antioxidant as well as by other mechanisms, can complement the direct effects induced by the intact ester. Taken together, these results suggest that unhydrolyzed vitamin E succinate, having a negative charge and stability in intracellular compartments, inhibits cell growth which may be in part due to the inhibition of PKC. Conceivably, structural modification of the dietary components will lead to the development of new pharmacological agents having greater chemopreventive potential while maintaining low toxicity.

Effects of Linoleic Acid on Breast Tumor Cell Proliferation Are Associated with Changes in P53 Protein Expression

J.K. TILLOTSON,[1] Z. DARZYNKIEWICZ,[2]
L.A. COHEN[1] AND Z. RONAI[1]

[1]*American Health Foundation, Valhalla, NY; and*
[2]*Cancer Research Center, New York Medical College,*
Valhalla, NY

A rat mammary tumor cell line maintained in low serum (1%) exhibited a slower growth rate, altered cell cycle distribution, decreased DNA synthesis, and increased immunoprecipitable p53, when compared with culture in 10% serum. Addition of linoleic acid (LA; 18:2, n-6) to low serum medium partially restored normal cell cycle distribution, increased synthesis of DNA, and decreased inmunoprecipitable p53 to levels normally seen in cells cultured in 10% serum. Recent experiments have indicated that LA is capable of regulating p53 in a similar fashion in human breast cancer cells. The capacity of LA to stimulate growth of these cell lines was inversely related to the cell's ability to activate transcription through a p53 recognition sequence. The inverse correlations between p53 levels, p53-mediated transcriptional activity, and cell proliferation are consistent with a role for p53 in negative regulation of growth in these tumor calls. The data suggest that growth stimulation of breast cancer cells by LA is mediated in part through modulation of p53 expression and activity.

Retinoic Acids and Retinoids in Mammary Tumor Prevention: Role of Supplemental Beta-Carotene and Vitamin A

E. FRIEDENTHAL, J. MENDECKI,
H. DAWSON and E. SEIFTER

Albert Einstein Medical College, Bronx, NY

Retinoic acids an._ derivatives, e.g. phenyl retinamides, are undergoing clinical evaluation as agents for preventing tumor recurrences following surgery, radiation or chemotherapy. In populations at high risk for developing mammary cancer, the role of these compounds will also be examined in prevention of first or second primary breast cancers.

The expectations for antineoplastic activity of retinoid compounds are based on the pro-epithelial differentative effects of the retinoids, their anti-proliferative action *vis-à-vis* mammary ductal tissue and to a lesser extent, their inflammatory effects. They will be studied under protocols similar to those used to evaluate Tamoxifen.

As we have demonstrated previously, retinoids possess certain activities characteristic of vitamin A (retinol); in fact, some of these activities become most apparent after their bioconversion to retinoic acid. However, serum retinoids regulate hepatic-derived serum retinol levels via a feedback mechanism, i.e. serum retinoids inhibit release of hepatic retinol and, as demonstrated by Underwood and ourselves, lower serum retinol. In doing so, they create a peripheral retinol deficiency.

In effect, therapeutic retinoids actually increase the requirements for beta-carotene and vitamin A, the very agents which are active in:
  a)   preventing metabolic stress responses due to therapies including retinoids
  b)   preventing oxidative stress due to radiation, hormones (both estrogens and anti-estrogens) and inflammation
  c)   enhancing thymotropic, lymphopoietic and immune reactions.
All of these act in tumor development. Moreover, retinol and retinal are required for sensory and transductive actions that influence breast tissue differentiation.

We believe that dietary supplementation of vitamin A and beta-carotene can maintain adequate blood and lymph levels of these nutrients even when retinoid therapy is employed.

Beta-Carotene Reduces Toxicity and Carcinogenicity of Cyclophosphamide in Control and Tumor Bearing Mice

J. MENDECKI, E. FRIEDENTHAL,
H. DAWSON and E. SEIFTER

Albert Einstein Medical College, Bronx, NY

Approximately 10% of cancer patients treated successfully with certain systemic alkylating or radiomimetic agents develop second malignancies related to their treatment. Similarly, a significant number of non-cancer patients with chronic inflammatory or immune dysfunctional diseases (i.e., psoriasis, arthritis) treated with systemic chemotherapy develop malignancies (mainly of lymphoid origin), while exposure to local irradiation tends to produce skin tumors.

Our previous work with beta-carotene and vitamin A showed these compounds to be effective in reducing the morbidity and mortality of total body and local irradiation in mice and of exposure to the radiomimetic agent streptozotocin.

In C_3H mice with transplanted C_3HBA mammary carcinoma, vitamin A and beta-carotene enhanced regression of the locally irradiated tumor and moderated toxic tumor or irradiation related metabolic responses. In similarly inoculated mice treated with cyclophosphamide (24 mg/kg 3 times a week) beta-carotene and vitamin A were both found to act synergistically with the chemotherapeutic agent to further inhibit tumor growth. While a higher dose of cyclophosphamide (36 mg/kg) gave better tumor regression and prolonged survival of a certain percentage of mice, it also caused many premature deaths due to its toxicity. However, animals receiving supplemental beta-carotene had a lower death rate in spite of the high dose of cyclophosphamide.

We previously reported that vitamin A decreased the incidence of lymphoma and leukemia in C57BL mice subjected to whole body irradiation (Kaplan model). We also found that vitamin A and beta-carotene each could reduce the rate of tumor induction in animals receiving chronic low doses of cyclophosphamide.

Beta-carotene and derived vitamin A may act by preventing oxidation of cyclophosphamide which causes production of carcinogens *in vivo*. We suggest that dietary beta-carotene and vitamin A may also enhance the synthesis of light dependent DNA repair enzymes previously known to be present in microorganisms and recently shown to operate also in mammals.

Does Boron Have a Special Role in Breast Cancer Detection and Treatment?

E. SEIFTER, J. MENDECKI,
H. DAWSON and E. FRIEDENTHAL

Albert Einstein College of Medicine, Bronx, NY

Boron is an essential nutrient for plants. Because of the previously known reactions of borates and of compounds having vicinal (not necessarily *cis*) hydroxyl groups, workers have found a role of borates in sugar transport. In diatoms boron is essential for bivalvular cell wall formation, indicating that borosilicates are intermediates of some skeletal biosyntheses. In women, supplemental boron also increases the anti-osteoporotic action of estrogen, vitamin D and supplemental calcium. Because of this action of boron, it is important to determine the status of boron as a dietary essential for animals. In animals borophosphates may play a role similar to that of borosilicates in diatoms. Even without such knowledge, it will be important to determine if supplemental boron concentrates in breast tissue and if selected boron compounds speed ductal calcification associated with some inflammatory and tumor processes. Earlier calcification may aid in earlier cancer detection.

Boron compounds in conjunction with neutron radiation have been employed in treatment of brain tumors. These treatments are based upon the lipophilicity of borates and their concentration by brain tissue. When boron containing tissues are exposed to neutrons, boron traps neutrons thermally and undergoes transmutation to lithium and releases carcino-lytic and carcinostatic levels of ionizing radiation. Borates may behave similarly in breast tissue where boron may concentrate due to the presence of catechol steroids and vicinal hydroxysteroids. Similarly borate esters of radioactive estrogens or vitamin D may be a convenient way of delivering radiation to breast tumor sites.

Contributors

Jan Bauer, Department of Oncology, Charles University, Prague, Czech Republic

William T. Cave, Jr., Endocrine Unit, Department of Internal Medicine, University of Rochester Medical Center, Rochester, New York 14642

Rowan T. Chlebowski, Division of Medical Oncology, Harbor-UCLA Medical Center, Torrance, California 90509

Sylvia Christakos, Department of Biochemistry and Molecular Biology, University of Medicine and Dentistry of New Jersey, New Jersey Medical School, Newark, New Jersey 07103

Carolyn K. Clifford, Diet and Cancer Branch, Division of Cancer Prevention and Control, National Cancer Institute, Bethesda, Maryland 20892

Jeanne M. Connolly, Division of Nutrition and Endocrinology, American Health Foundation, Valhalla, New York 10595

Nikolay V. Dimitrov, Department of Medicine, Michigan State University, East Lansing, Michigan 48824

Kent L. Erickson, Department of Cell Biology and Human Anatomy, University of California School of Medicine–Davis, Davis, California 95616

Laurence S. Freedman, Biometry Branch, Division of Cancer Prevention and Control, National Cancer Institute, Bethesda, Maryland 20892

Barry R. Goldin, Department of Community Health, Tufts University School of Medicine, Boston, Massachusetts 02111

Sherwood L. Gorbach, Department of Community Medicine, Tufts University School of Medicine, Boston, Massachusetts 02111

Neil E. Hubbard, Department of Cell Biology and Human Anatomy, University of California School of Medicine-Davis, Davis, California 95616

Clement Ip, Department of Surgical Oncology, Roswell Park Cancer Institute, Buffalo, New York 14263

Thomas I. Jones, Department of Medicine, Michigan State University, East Lansing, Michigan 48824

Xin-Hua Liu, Division of Nutrition and Endocrinology, American Health Foundation, Valhalla, New York 10595

Richard C. Moon, Specialized Cancer Center, University of Illinois at Chicago, Chicago, Illinois 60612

Harold L. Newmark, Memorial Sloan-Kettering Cancer Center, New York, New York 10021

Rui-Qin Pan, Department of Surgery, China-Japan Friendship Hospital, Beijing, China

Ross L. Prentice, Division of Public Health Sciences, Fred Hutchinson Cancer Research Center, 1124 Columbia Street, Seattle, Washington 98104

David P. Rose, Division of Nutrition and Endocrinology, American Health Foundation, Valhalla, New York 10595

Joseph A. Scimeca, Nutrition Department, Kraft General Foods, Glenview, Illinois 60025

Henry J. Thompson, Laboratory of Nutrition Research, AMC Cancer Research Center, Denver, Colorado 80214

Index